MEDICAL
INTELLIGENCE
UNIT

DNA Repair
and Human Disease

Adayabalam S. Balajee, M. Phil., Ph.D.

Center for Radiological Research
Department of Radiation Oncology
College of Physicians and Surgeons
Columbia University
New York, New York, U.S.A.

LANDES BIOSCIENCE / EUREKAH.COM
GEORGETOWN, TEXAS
U.S.A.

SPRINGER SCIENCE+BUSINESS MEDIA
NEW YORK, NEW YORK
U.S.A.

DNA Repair and Human Disease

Medical Intelligence Unit

Landes Bioscience / Eurekah.com
Springer Science+Business Media, LLC

ISBN: 0-387-34195-1   Printed on acid-free paper.

Springer Science+Business Media, LLC, 233 Spring Street, New York, New York 10013, U.S.A.
http://www.springer.com

Please address all inquiries to the Publishers:
Landes Bioscience / Eurekah.com, 810 South Church Street, Georgetown, Texas 78626, U.S.A.
Phone: 512/ 863 7762; FAX: 512/ 863 0081
http://www.eurekah.com
http://www.landesbioscience.com

Printed in the United States of America.

9 8 7 6 5 4 3 2 1

Library of Congress Cataloging-in-Publication Data

DNA repair and human disease / [edited by] Adayabalam S. Balajee.
    p. ; cm. -- (Medical intelligence unit)
  Includes bibliographical references and index.
  ISBN 0-387-34195-1 (alk. paper)
  1. DNA repair. 2. Medical genetics.   I. Balajee, Adayabalam S.
  II. Series: Medical intelligence unit (Unnumbered : 2003)
  [DNLM: 1. DNA Repair--physiology. 2. Disease--etiology.
QU 475 D6288 2006]
QH467.D163 2006
616'.042--dc22
                                    2006012699

*This book is dedicated
to my father, Adayapalam Thyagarajan Sambasivan
and to my beloved family members.*

=CONTENTS=

EDITOR

Adayabalam S. Balajee
Center for Radiological Research
Department of Radiation Oncology
College of Physicians and Surgeons
Columbia University
New York, New York, U.S.A.
Email: ab836@columbia.edu
Chapter 7

CONTRIBUTORS

Byungchan Ahn
Department of Life Sciences
University of Ulsan
Ulsan, Korea
Chapter 1

Colette apRhys
Mckusick-Nathan Institute
    of Genetic Medicine
Howard Hughes Medical Institute
Johns Hopkins University
    School of Medicine
Baltimore, Maryland, U.S.A.
Email: Colette@jhmi.edu
Chapter 10

Alan Ashworth
Breakthrough Breast Cancer Centre
Institute of Cancer Research
London, U.K.
Email: alana@icr.ac.uk
Chapter 4

Massimo Bogliolo
Department of Genetics
    and Microbiology
Mutagenesis Group
Universitat Autònoma de Barcelona
Barcelona, Spain
Chapter 6

Vilhelm A. Bohr
Laboratory of Molecular Gerontology
National Institutes of Health
National Institute on Aging
Baltimore, Maryland, U.S.A.
Email: vbohr@nih.gov
Chapter 1

Marcus S. Cooke
Department of Cancer Studies
    and Molecular Medicine
and
Department of Genetics
University of Leicester
Leicester, U.K.
Chapter 8

Charles R. Geard
Department of Radiation Oncology
Center for Radiological Research
College of Physicians and Surgeons
Columbia University
New York, New York, U.S.A.
Chapter 7

Katrin Gudmundsdottir
Breakthrough Breast Cancer Centre
Institute of Cancer Research
London, U.K.
Chapter 4

M. Prakash Hande
Department of Physiology
National University of Singapore
Singapore
Email: phsmph@nus.edu.sg
Chapter 9

Daniel Judge
Department of Medicine
Division of Cardiology
Johns Hopkins Hospital
Baltimore, Maryland, U.S.A.
Chapter 10

Takehisa Matsumoto
Genome Center
Japanese Foundation for Cancer Research
Tokyo, Japan
Email: takehisa.matsumoto@jfcr.or.jp
Chapter 2

Adayapalam T. Natarajan
Department of Radiation Genetics
   and Chemical Mutagenesis
State University of Leiden
Leiden, The Netherlands
Email: Natarajan@xs4all.nl
Chapter 5

George Perry
Institute of Pathology
Case Western Reserve University
Cleveland, Ohio, U.S.A.
Chapter 8

V. Prakash Reddy
Department of Chemistry
University of Missouri-Rolla
Rolla, Missouri, U.S.A.
Email: preddy@umr.edu
Chapter 8

Lawrence M. Sayre
Department of Chemistry
Case Western Reserve University
Cleveland, Ohio, U.S.A.
Chapter 8

Mark A. Smith
Institute of Pathology
Case Western Reserve University
Cleveland, Ohio, U.S.A.
Email: mas21@po.cwru.edu
Chapter 8

Miria Stefanini
Instituto di Genetica Molecolare
Consiglio Nazionale delle Ricerche
Pavia, Italy
Email: stefanini@igbe.pv.cnr.it
Chapter 3

Jordi Surrallés
Department of Genetics
   and Microbiology
Mutagenesis Group
Universitat Autonoma
   de Barcelona Bellaterra
Barcelona, Spain
Email: jordi.surralles@uab.es
Chapter 6

Emily Witt
Breakthrough Breast Cancer Centre
Institute of Cancer Research
London, U.K.
Chapter 4

PREFACE

Endogenous and exogenous agents, which induce various types of DNA lesions ranging from simple base alterations to complex helix distorting bulky DNA adducts, constantly threaten the integrity of genomic DNA. These DNA lesions depending upon the type and specificity of induction differentially affect the fidelity of DNA replication and transcription leading to mutations in important protein coding gene sequences. The mutated proteins in turn affect the various biological processes leading to an increase in the incidence of carcinogenesis and genomic instability. Such changes, if they occur in germ cells, are heritable and so it is important to protect the DNA from damage and maintain the genetic information contained in the DNA. Prokaryotic and eukaryotic organisms are equipped with diverse DNA repair pathways to protect their DNA. The importance of DNA repair pathways in the maintenance of genomic stability is best illustrated by a wide array of naturally occurring human autosomal recessive disorders, which exhibit deficiencies in one or more pathways of DNA repair. DNA repair, rightfully earning the importance of being the molecule of the year in 1994, has greatly integrated ever since into different disciplines of scientific inquiry including the molecular basis of human health, cancer and aging. Recent advances in unraveling the fascinating molecular link between DNA repair deficiency and human diseases has actually prompted me to provide the scientific community with a thorough state of the art of information on the importance of DNA repair in genomic integrity. This book comprises a collection of some of the important human disorders with a comprehensive description of clinical manifestations as well as the cellular and molecular aspects of the genes responsible for the diseases. Further, each chapter is provided with a latest collection of references that might be useful for biologists interested in these various human syndromes. The book is intended to be useful for students at all levels and researchers working in the areas of cell biology, biochemistry, molecular biology and carcinogenesis.

Aging is an unavoidable biological process, and recent studies have suggested an important role for DNA repair in aging process. The book starts with a description of a molecular connection between premature aging and DNA repair dysfunction in Werner (WRN) and Cockayne syndromes (CS). This chapter brings out a detailed description of the molecular functions of WRN and CS group B genes and how deficiencies in these genes predispose human individuals to some of the accelerated aging features. This chapter is followed by DNA repair aspects in some of the RecQ family of helicase disorders namely Werner and Bloom syndromes. Chapter 3 focuses on the intrinsic connection between DNA repair and transcription and illustrates how deficiencies in these overlapping functions can lead to an autosomal human disorder, trichothiodystrophy. The fourth chapter deals with the role of breast cancer susceptibility (BRCA1 and BRCA2) genes in DNA repair and documents the importance of DNA repair in breast cancer incidence. Suggestive evidences for linking the deficiency in the repair of ionizing radiation induced DNA strand breaks with the pathogenesis of Down syndrome are provided in Chapter 5. The sixth chapter provides a comprehensive account of the importance of Fanconi anemia and BRCA pathway in the maintenance of genomic stability. Evidence for the molecular link between DNA repair and cell cycle checkpoint regulators in the pathogenesis of extremely radiosensitive human ataxia telangiectasia patients is provided in Chapter 7. The molecular basis of neurodegeneration in Alzheimer's disease with special emphasis on oxidative DNA damage and repair is presented in the eighth chapter. This is followed by a chapter that highlights the importance of co-ordination between telomeres and DNA repair factors in the maintenance of chromosome-genome integrity by giving suitable illustrations using knockout mouse model systems. The final chapter forcefully brings out the importance of xeroderma pigmentosum and Cockayne syndrome genes in the avoidance of sunlight-induced skin cancer in humans.

This book would not have been possible without the dedicated involvement of all the authors and co-authors who have generously contributed to these chapters. I deeply express my sincere gratitude to each one of them for their contribution. I am also grateful to R.G. Landes and his publishing staff, especially Ms. Cynthia Conomos, who have relentlessly worked hard for the development and outcome of this book. I sincerely hope this book will stimulate novel ideas among scientists leading to a better understanding of the molecular basis of human health, aging and cancer.

Adayabalam S. Balajee, M.Phil., Ph.D.

CHAPTER 1

Human Premature Aging Disorders and Dysfunction of DNA Repair

Byungchan Ahn and Vilhelm A. Bohr

Abstract

Werner's syndrome (WS) and Cockayne syndrome (CS) are rare human autosomal recessive disorders classified as segmental progeroid disorders. WS is marked by premature onset of age-related phenotypic changes (such as cataract and greying of hair etc) and genome instability. Cells derived from CS patients are defective in DNA repair, and CS patients display severe neurological abnormalities and certain features of premature aging. The accelerated aging features observed in these two genetically distinct syndromes make them ideal model systems to understand some of the basic mechanism(s) for aging process. WS is caused by mutations in the gene (wrn) encoding the Werner syndrome protein (WRN), a member of the RecQ family of DNA helicases with helicase and exonuclease domains. Biochemical characterization of WRN enzymatic activities, identification of WRN interacting proteins and the cellular localization studies all suggest WRN's participation in multiple DNA metabolic pathways for maintaining genomic integrity. CS arises from mutations in the CSA and CSB genes. While CSA belongs to "WD" repeat containing protein family, CSB is a member of SNF2-like family with only a DNA stimulated ATPase activity. CSB is involved in different DNA transactions, notably transcription and DNA repair. Our recent understanding of the mechanism of action(s) of WRN and CSB proteins are expected to provide new insights into the aging process.

Introduction

The study of heritable genetic diseases, which are characterized by accelerated onset of some of the normal aging features, provides insights into how the aging process occurs in normal individuals. A number of progeroid syndromes have been described in humans. In patients with these disorders, aging-like symptoms and age-associated diseases appear much earlier than the age matched normal individuals. Therefore, progeroid syndromes are considered an excellent model for the study of the aging process.

Werner syndrome (WS) is classified as an adult progeria and patients with WS have very significant clinical presentation of natural aging, such as graying of the hair and arteriosclerosis. At the cellular level, the clearest features of WS cells include extremely poor replicative potential and variegated translocation mosaicism. The gene (WRN) mutated in this syndrome has been identified; WRN is classified as a RecQ helicase and interestingly encodes an exonuclease activity.

DNA Repair and Human Disease, edited by Adayabalam S. Balajee. ©2006 Landes Bioscience and Springer Science+Business Media.

Patients with Cockayne syndrome (CS) are characterized by traits reminiscent of normal aging and CS is also categorized as a progeroid syndrome. Two genes, CSA and CSB, were identified and the majority of CS cases are due to mutations in the CSB gene. CSB protein has a DNA-dependent ATPase activity but no measurable helicase activity in vitro in spite of conserved helicase motifs. CS-B cells are specifically defective in both transcription-coupled nucleotide excision repair and global base excision DNA repair pathways.

It is intriguing to speculate how a single mutated gene can lead to diverse phenotypic changes associated with human aging. This can be partly explained by the interactions of WRN and CSB with proteins involved in pathways of DNA metabolism.

Werner Syndrome

WS, named after the founder Dr. Otto Werner in 1904, was first documented with two principal symptoms: bilateral cataracts and scleoderma-like skin.[152] WS is an autosomal recessive disorder manifested by an early onset of aged-appearance and age-related common disorders.[69] Individuals with the *WRN* null mutations display a remarkable number of clinical signs and symptoms associated with aging early in their life (premature aging). WS patients have a normal development until 20 years of age. However, after the adolescent growth spurt, patients develop graying of the hair, hair loss, cataracts, dermal atrophy, diabetes mellitus (type II), osteoporosis, atherosclerosis and cancers.[69,104] In support of WS as a model system of normal aging, a recent extensive DNA micro-array analysis of WS and normal, young and old fibroblasts showed that the gene expression pattern of WS is very similar to that of normal aging.[60]

WS has been also described as segmental progeroid disorder because the disease displays some, but not all of the symptoms of aging. For example, WS patients usually do not develop Alzheimer-type dementia. In addition, WS is characterized by a number of clinical features unrelated to normal aging, including calcification, hypogonadism, short stature, skin ulcers and reduced fertility.

Cellular Phenotypes of WS

A number of cellular defects associated with WS have been characterized. One of the major hallmarks is genome instability. WS cells have a wide range of chromosomal abnormalities from chromosome deletions to translocation and complex rearrangements. An interesting cytogenetic feature of WS cells is a variegated translocation mosaicism (VTM).[105] This mosaicism is defined as the expansion of variable, clonal chromosome rearrangements in different clones from the same cell line.[47] Although WS fibroblasts display accelerated rates of telomere shortening,[58,109] drastic shortening or lengthening of the telomere lenghts have been observed during cell passage.[130] Along with the chromosomal instability, WS cells exhibit a reduced replicative life span,[70,106] a delayed S phase,[97] a reduced frequency of initiation sites,[42,131] and a hypersensitivity to agents such as cross-linkers and topoisomerase inhibitors that interfere with DNA replication.[64,94,96,98] WS cells also display reduced repair of psoralen cross-links[8] and reduced repair of DNA damage induced in S-phase cells. In addition to replication defects, WS cells show hypersensitivity to the chemical carcinogen 4-nitroquinoline-1-oxide (4-NQO)[82,95] and mild sensitivity to ionizing radiation[102,157] (Fig. 1). However, WS cells are not particularly sensitive to other DNA damaging agents such as alkylating agents, hydroxyurea, bleomycin, UV-irradiation and X-rays[34,36,83,123] indicating that the repair deficiency of WS cells is selective only to certain types of DNA lesions. By transfecting cells with linearized plasmids or shuttle vectors, WS cells showed extensive degradation of DNA ends before ligation, particularly at the 5'-recessed ends.[89] Collectively, these observations suggest a specialized DNA repair defect in WS. In addition, WS displays a reduced RNA polymerase II transcription.[2] WS lymphoblast cell lines carrying homozygous

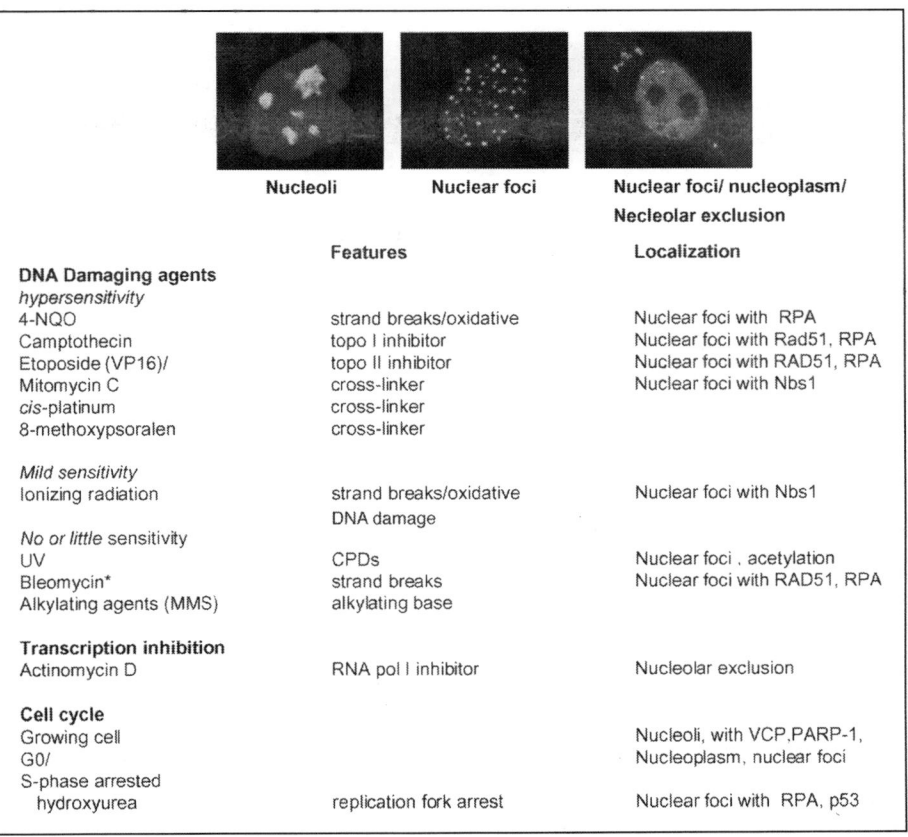

	Features	Localization
DNA Damaging agents		
hypersensitivity		
4-NQO	strand breaks/oxidative	Nuclear foci with RPA
Camptothecin	topo I inhibitor	Nuclear foci with Rad51, RPA
Etoposide (VP16)/	topo II inhibitor	Nuclear foci with RAD51, RPA
Mitomycin C	cross-linker	Nuclear foci with Nbs1
cis-platinum	cross-linker	
8-methoxypsoralen	cross-linker	
Mild sensitivity		
Ionizing radiation	strand breaks/oxidative	Nuclear foci with Nbs1
	DNA damage	
No or little sensitivity		
UV	CPDs	Nuclear foci , acetylation
Bleomycin*	strand breaks	Nuclear foci with RAD51, RPA
Alkylating agents (MMS)	alkylating base	
Transcription inhibition		
Actinomycin D	RNA pol I inhibitor	Nucleolar exclusion
Cell cycle		
Growing cell		Nucleoli, with VCP,PARP-1,
G0/		Nucleoplasm, nuclear foci
S-phase arrested		
hydroxyurea	replication fork arrest	Nuclear foci with RPA, p53

Figure 1. Sub-cellular localization of the Werner protein. WRN is localized differently in response to DNA damaging agents, transcription inhibitor and cell-cycle stages.

mutations in the WRN gene showed 40-60% of the transcription activity observed in normal cells using a [³H]-uridine incorporation assay with cells, or an in vitro assay with nuclear extracts and a RNA pol II specific template.[2] When purified wild-type WRN protein was added to the nuclear extracts, RNA pol II transcription was markedly stimulated, suggesting that a transcription defect in WS cells is due to the absence of function of WRN protein. A role of WRN in transcription is also supported by enhanced p53-dependent transcriptional activity by WRN over expression.[6] In addition, p53-mediated apoptosis is attenuated in the absence of WRN.[122]

Thus, WRN-dependent activation of the p53 response may prevent the proliferation of cells with damaged genomes and WRN may participate in the activation of p53 in response to certain types of DNA damage. Understanding the mechanism by which WRN participates in the DNA repair pathways may provide insight into the genetic instability and the accelerated aging in WS.

The WRN Gene Product and Its Biochemical Activities

The *WRN* gene located on the short arm of chromosome 8 consists of 35 exons encoding a protein of 1432 amino acids. It was identified by positional cloning in 1996.[158]

WRN interaction partners

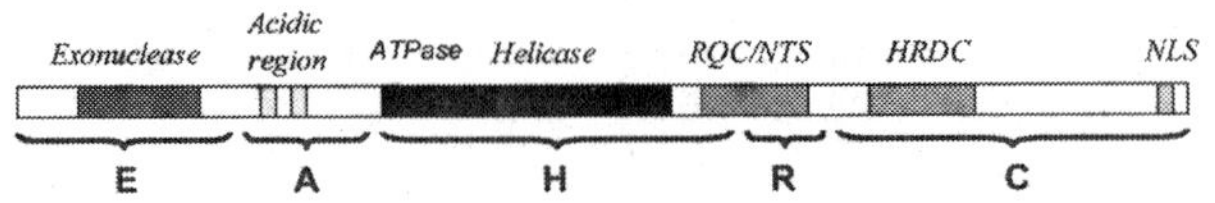

WRN Domain	Partner	WRN HEL	WRN EXO	Partner activity	Ref
C-ter	RPA	increase			Shen et al., 1998;Brosh et al., 1999
C-ter(R)	TRF2	increase			Opresko et al, 2002
N-ter&H	Nbs1	increase			Cheng et al., in press
C-ter	Rad52	Increase/ inhibit			Baynton, 2003
C-ter	Ku70/80		increase		Cooper at al.,2000; Li et al.,2000;Karmakar et al., 2002
N-ter	Ku80		increase		Shen et al.,2003
C-ter(R)	p53		inhibit		Blander et al.,1999;Spillare et al.,1999
N-ter(E)/H	BLM		inhibit		Von Kobbe et al.,2002
	DNA-PKcs	Inhibit/p	Inhibit/p		Yannone et al.,2001;Karmakar et al.,2002; Li,2002
N-ter&H	C-Abl	Inhibit/p	Inhibit/p		Cheng et al.,2003
N-ter(E)	PCNA				Lebel et al.,1999
C-ter(R)	Pol d			stimulate	Szekely at al., 2000
C-ter	Pol b			stimulate	Harrigan et al.,2003
C-ter(R)	FEN-1			stimulate	Brosh et al.,2001, Sharma et al.,2004
C-ter(R)	PARP-1				Von Kobbe et al.,2003, Li et al.,2004
C-ter(R)	PKA				Nguyen et al.,2002
	Topo I			stimulate	Laine et al.,2003
	VCP				Partridge et al.,2003

Figure 2. Proteins interacting with WRN. Mostly, the C-terminal and N-terminal domain of WRN interacts physically with many proteins and the interactions are functionally related.

The *WRN* gene product is a member of the RecQ DNA helicase family with high sequence homologies to the *E. coli* RecQ, the F18C5C gene of *C. elegans*, yeast SGS1 and RQH1, human RECQL, RECQ4, RECQ5, and BLM proteins, and *Xenopus* FFA-1.[57,59] Structural analysis of WRN showed four defined regions (see Fig. 2). An exonuclease domain is localized at the N-terminus of the protein[77,79] and the RecQ-type helicase domain is located at a central domain of ~600 residues comprising the seven conserved motifs.[38] A RecQ conserved domain is located C-terminal to the helicase domain.[76] Finally, a helicase ribonuclease D C-terminal (HRDC) domain found in a number of DNA metabolizing proteins is at the C-terminal region.[76] Additionally, a transcriptional activation domain, a nuclear localization signal, and nucleolus targeting sequences all have been identified by cellular studies. Biochemical studies confirmed the helicase and exonuclease activities of WRN in vitro.[39,50,125] Using recombinant truncated fragments of WRN, three DNA binding domains were identified including RQC domain, HDRC domain, and one N-terminal fragment.[148]

WRN Helicase Activity

Initial studies showed that WRN helicase unwinds relatively short DNA duplexes of ≤ 53 bp depending on the presence of ATP and divalent cations, with a 3'-5' directionality.[39,125] While WRN effectively displaces shorter oligonucletides it is relatively less efficient in unwinding longer substrates such as a 257 bp duplex DNA,[14] indicating a limited processivity in displacement. However, in the presence of the single-stranded DNA binding protein Replication Protein A, WRN helicase can catalyze unwinding of long duplexes up to 851 bp.[14] The mechanism by which RPA enhances the WRN helicase activity probably involves the stabilization of displaced single-stranded DNA by preventing the reannealing process.

In addition to DNA duplex length properties of WRN, WRN helicase unwinds a variety of DNA substrates mimicking the intermediates of DNA replication, recombination, and repair in vitro. For example, WRN unwinds RNA-DNA heteroduplexes,[125] DNA triplexes,[13] flap, G4 tetraplexes that made by two hairpin loop structures[33,75] as well as D-loop that resembles telomere ends.[86a,87] Furthermore, WRN unwinds a blunt-ended duplex containing a 12 nt single stranded bubble and a synthetic Holliday junction recombination intermediate.[75,112] In fact, secondary DNA structures found throughout the genome can potentially interfere with cellular processes (DNA replication, transcription and recombination) and may result in genomic instability. The capability of WRN to unwind these substrates therefore supports a role for WRN in many aspects of DNA metabolism and some of the defects in DNA replication, transcription and recombination detected in WS cells may be a consequence of the absence of WRN helicase activity to resolve these structures.

WRN Exonuclease Activity

WRN possesses an exonuclease activity.[50,53,113] The presence of a unique exonuclease domain located at the N-terminal region clearly delineates WRN from the other RecQ helicases of human, yeast and *E. coli*. This exonuclease domain is homologous to the exonuclease of *E. coli* DNA polymerase I and RNaseD.[77] Four independent biochemical characterizations of WRN have demonstrated that it is a 3'-5' exonuclease activity.[50,53,113] However, one group has reported a 5'-3' exonuclease activity.[126] The WRN exonuclease initiates degradation from 3'-recessed termini of duplex DNA, and is also able to initiate digestion at a blunt end if the substrate contains a junction or alternate structures such as a bubble, a fork, and a Holliday junction.[85,112,116] However, WRN does not digest blunt ended, recessed 5'OH-bearing, and ssDNA substrates.

It had been reported that the exonuclease and helicase activities of WRN are physically and functionally separable.[50] In addition, a recombinant N-terminal fragment for WRN that lacks the ATPase/helicase domain retains exonuclease activity,[24,50] indicating that exonuclease activity can be dissociated from the helicase activity. Nevertheless, the exonuclease activity of the full-length recombinant WRN protein is stimulated by the presence of ATP.[116] In addition, it was recently shown that a forked duplex substrate containing one blunt end could be simultaneously acted upon at the opposite ends by WRN helicase and exonuclease,[85] indicating that the exonuclease activity may function coordinately with ATP binding/hydrolysis to process DNA metabolic intermediates. If WRN acts as a monomer, the space between the helicase domain and the exonuclease domain of a monomer WRN may affect the coordination. Thus, it remains to be determined how duplex lengths of forked duplex substrates affect the exonuclease activity. If WRN acts as a multimer, the intermolecular interaction of WRN and the length between the active helicase and exonuclease domain of the multimeric WRN complex may be important factors.

Cellular Location of WRN

WRN has a NLS (nuclear localization signal) near the C-terminus of the protein (amino acids 1370-1375) and has been detected in both the nucleoplasm and nucleolus (Fig. 1).[71] In addition, a nucleolar localization sequence has been determined.[146] The distribution of the WRN molecules located in different nuclear compartments is variable and is obviously regulated by several endogenous factors such as the cell cycle distribution, post translational modifications, and exogenous factors like different fixation protocols employed for WRN detection. Most mutations in the WS patients result in a truncated WRN protein that lacks the NLS (reviewed in ref. 88). Hence, inability of WRN to transport into the nucleus is likely to be critical for the pathogenesis of WS.

The majority of WRN localizes to the nucleolus in exponentially growing cells, whereas WRN was found outside the nucleoli in quiescent[37] and S-phase arrested cells.[11,23] When human cells were treated with 4-NQO, UV, bleomycin, serum starvation and etoposide, WRN was translocated from the nucleolus to the nucleoplasm after treatment.[6,11,40,103,127] Furthermore, WRN leaves the nucleoli upon inhibition of rRNA transcription following actinomycin D treatment.[117]

The nucleolar localization is somehow modulated by post-translational change of WRN. Recent work indicates that acetylation of WRN by p300 augments translocation of the protein to nuclear foci.[6] Very recently, it was shown that WRN and p97/VCP (vasolin-containing protein) physically interacted in the nucleoli.[91] Dissociation of VCP and WRN from each other and the subsequent relocation of WRN from the nucleolus after CPT treatment suggest that the change in WRN localization is by post-translational modification.[91] This evidence implies that VCP may play a role in the response to DNA damage by modulating the nucleolar trafficking of WRN. However, the consequences of these modifications in the transport of WRN into and out of the nucleoli and the role of WRN in the nucleolus remain to be elucidated.

The nucleolar targeting of the WRN protein is due the presence of a nucleolar targeting sequence (NTS) in the C-terminal of the protein.[146] An extensive mapping study showed that the WRN NTS spans a 144 aa region (949-1092) of the RecQ conserved (RQC) domain. This NTS is dependent on the presence of an active NLS, indicating that the nucleolar targeting process is directly linked to the nuclear import process. In contrast, one group proposed that the basic C-terminal aa (1403-1404) represents the WRN NTS.[127] The reason for this discrepancy is presently unknown but can be attributed to variations in the experimental procedures.

Cellular Function of WRN

Cellular studies of WS cells together with the identification of WRN interacting proteins have shed light into the in vivo functions of WRN (Fig. 2). These findings suggest that WRN has multiple functions through its involvement in a wide variety of DNA metabolic processes (Fig. 3). Most of the proteins interacting physically with WRN show functional interactions. Several approaches have been undertaken to characterize the interacting proteins: binding to specific candidates, screening cDNA libraries by a yeast two-hybrid system, isolating the WRN complex from cell extract, using a WRN affinity column;[5,129] for details, see a recent review.[84] Collectively, these approaches have identified a number of proteins that interact with WRN. WRN may interact differentially with different proteins depending on the status of DNA metabolism in which WRN is involved during the time of the experiments. It is therefore important to distinguish between the proteins that bind to WRN with or without functional consequence(s). We have obviously ascribed more significance to those protein interactions that are associated with a functional consequence, judged by a change in WRN helicase or exonuclease activity by the interacting protein in question. This aspect was reviewed in ref. 84 and the protein interactions of functional consequence are shown in Figure 3 in bold.

Recombination

WS cells have several phenotypes that are suggestive of genome instability. These include reduced cell proliferation,[70,106] sensitivity to DNA damaging agents such as 4NQO, cross-linking agents and camptothecin,[82,95,96] cytogenetic abnormalities,[74,105] and a deletion mutator.[35] This homologous recombination (HR) pathway protects the cells from double-strand breaks (DSBs) generated by exogenous agents including ionizing radiation, DNA cross-linkers and stalled or inactivated replication forks. Failure to repair the DSBs can lead to chromosome breakage or loss, gene rearrangement and mutation or cell death.[66,72] Increased chromosomal instability in

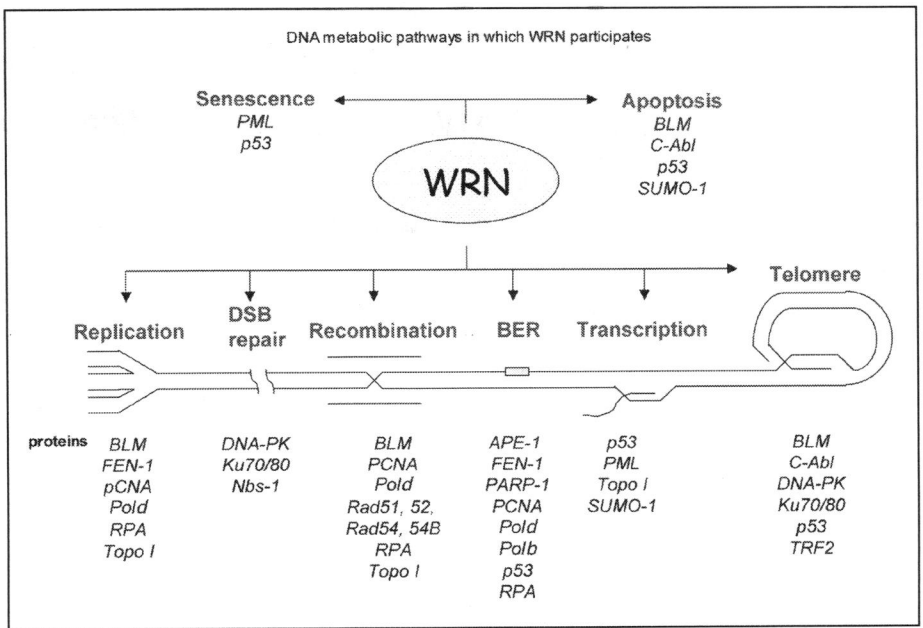

Figure 3. DNA metabolic pathways in which WRN may participate. The WRN interacting partners are involved in several DNA metabolisms which lead us to postulate possible pathways of WRN's particpation.

the form of VTM and loss of large genomic fragments in WS cells suggest the involvement of WRN in the homologous recombinational repair pathway to repair DSBs and to maintain chromosomal stability.

Since the *WRN* gene is a homolog of *E. coli* RecQ[139] and *S. cerevisiae* Sgs1 genes,[68,151] genetic studies from these homologues have enabled us to speculate on the in vivo function of WRN in humans. Studies in *E. coli* showed that RecQ helicase is involved in homologous and illegitimate recombination.[43] In yeast, defects in *SGS1* lead to higher recombination rates.[80,156] In addition, the *sgs1* mutation reduces the average lifespan of yeast cells.[73,118] Strikingly, a WRN helicase activity can suppress the increased homologous and illegitimate recombination in the *S. cerevisiae sgs1* mutant.[156] These findings imply a role for the WRN helicase activity in homologous recombination.

Recent studies revealed a marked reduction in cell proliferation following mitotic recombination, and a few viable gene conversion-type recombinants in WS cells.[99] However, recombination initiation and rates occurred at comparable rates in WS cells and control cell lines,[102] indicating that other steps in recombination would be defective in WS cells.

To elucidate the WRN recombination defect and the inter-relationship between HR and cell survival following DNA damage, either wild-type WRN protein or the bacterial resolvase protein RusA was expressed in WS cells. Remarkably, WS cell survival and the generation of viable mitotic recombinant progeny was rescued after DNA damage.[102] These results implicate a deficiency in resolving recombination intermediates in the absence of functional WRN. In addition, the dependence of WRN cellular phenotypes on RAD51-dependent HR pathways was investigated by using a dominant-negative RAD51 protein.[102] The DNA damage hypersensitivity to *cis*-platinum and hydroxyurea in WS cells was reversed. This result indicates that HR function is important for the generation of WRN cellular phenotypes.

WRN may mediate the resolution of recombinational intermediates in association with RAD51, RPA, BLM and possibly yet unidentified protein partners.[14,12] In support of this, a recent report shows that WRN colocalizes with RAD51 and RPA in response to CPT treatment[103] specifically in S-phase cells. Very recently, WRN was found to interact physically and functionally with Mre11/Rad50/Nbs1 (MRN) complex and to colocalize with the complex in response to ionizing radiation or mitomycin C treatment.[18] Since the MRN complex functions in HR, the mechanism of this interaction is being explored. Understanding the roles of WRN in recombinational repair is at an exciting but early stage.

DNA Replication

Several lines of evidence point to a DNA replication defect in WS cells: (i) premature replicative senescence,[70,106] (ii) a prolonged S-phase,[97] and (iii) an altered frequency of DNA replication initiation.[42,131] This evidence is in support of a direct role for the WRN protein in DNA replication. In addition, WRN has been found to physically and functionally interact with several components of the DNA replication complex, including RPA, PCNA and topoisomerase I (topo I).[14,61,65,114,115] Since RPA has demonstrated roles in DNA replication, recombination, and repair, WRN also is likely to function with RPA in one or more of these processes. The N-terminal domain of the large subunit of RPA interacts with the N-terminal region of WRN and RPA enables the WRN helicase to unwind long duplexes,[14,115] suggesting that WRN helicase may function in some capacity as the replication fork progresses. Recently, it has become clear that recombination plays an important role in the repair of stalled or broken replication forks, leading to the reinitiation of replication.[41,100] Mutants of *S. cerevisiae sgs1*, a yeast homologue of WRN, display a hyper-recombination phenotype,[80,101] which can be partially suppressed by WRN expression.[156] WRN colocalizes with RPA upon replication fork arrest after treatment of cells with hydroxyurea, and promotes translocation of Holliday junctions.[23] Furthermore, WRN unwinds efficiently the chicken-foot Holliday Junction intermediate associated with a regressed replication fork,[112] Thus, these lines of evidence support a role of WRN for resolving the block and/or participating in the reinitiation of replication. In addition, it has been shown that WS cells are hypersensitive to the topo I inhibitor and that WRN stimulates topoI relaxation activity in vitro.[61,93] Therefore, WRN may act with topo I during replication to resolve blocks caused by topological problems in the DNA.

Studies in yeast indicate an important role for FEN-1 in the processing of Okazaki fragments,[4] and biochemical and genetic evidences indicate that the double flap structure generated during DNA synthesis strand displacement is the physiological substrate of the enzyme.[54,155] WRN was found to interact with FEN-1 by coimmunoprecipitation, and to stimulate the flap cleavage activity of FEN-1 in vitro. Thus, the interaction of WRN with FEN-1 implicates a role of WRN in the processing of Okazaki fragments during lagging DNA synthesis. In support of this idea, WRN and FEN-1 form a complex in vivo that colocalize in foci associated with arrested replication forks.[112] Biochemical analyses showed that WRN helicase unwinds the chicken-foot Holliday Junction intermediate generated at a regressed replication fork and stimulate FEN-1 to cleave the unwound strand in a structuredependent manner. Thus, WRN and FEN-1 function together to process replication fork structures.[112]

Base Excision Repair

Hypersensitivity of WS cells to oxidative damage generating DNA damaging agents (4-NQO, hydrogen peroxide) suggests that WRN may participate in some aspects of repair of oxidative DNA damage.[82,95,147] Since aging phenotypic changes are hypothesized to be due to the accumulation of oxidative DNA products, the involvement of WRN in base excision repair (BER) is of importance in explaining the premature appearance of normal aging in WS patients.

Extracts from WS cells have been shown to efficiently repair a plasmid containing an abasic (AP) site, which is generally repaired by short-patch BER, indicating that WS cells are proficient in this repair event. However, the ability of WS cell extracts to repair lesions requiring long-patch BER in vitro remains to be investigated. If WS cells are deficient in long patch BER, it will be important to find DNA-damaging agents that generate DNA substrates specific for long-patch BER and then test their effects on WS cells.

Recently, in vitro biochemical studies suggest that WRN interacts physically and/or functionally with several long-patch BER proteins, pol δ, PCNA, RPA, and FEN-1 (see Fig. 2). WRN interacts physically with FEN-1 and stimulates its DNA flap cleavage activity.[10,15] As deficiencies in the removal of the 5' dRP-containing flap could potentially lead to an undesirable ligation at an AP-nick or a strand break in the opposite strand, the stimulation of FEN-1 activity by WRN suggests the participation of WRN in long-patch BER. Very recently a physical and functional interaction between WRN and pol β was identified.[44] An active WRN helicase domain stimulates pol β strand displacement DNA synthesis at a nick on a BER intermediate.[44] However, such stimulation was not observed on a primer/template substrate lacking a downstream oligonucleotide.[44] Therefore, WRN stimulation of pol β appears to be dependent on DNA substrates. In addition, WRN can unwind a BER substrate produced following uracil-DNA glycosylase and AP endonuclease (APE1) treatment of a uracil-containing oligonucleotide,[44] suggesting that a displacement of the downstream strand may be important for long-patch BER. Together, these results provide further evidence for a role of WRN in BER.

PARP-1 (poly(ADP-ribose) polymerase I) is a nuclear enzyme involved in DNA replication and repair. PARP-1 is activated by several DNA damaging-agents, such as hydrogen peroxide, alkylating agents (like MMS), bleomycin, and radiation.[16,149] The ADP-ribosylation activity thus appears to be important for responding to DNA damage and perhaps facilitating DNA repair, thereby maintaining genome stability. Very recently, it was identified that PARP-1 interacts strongly with WRN.[1,147] In WS cells, the poly(ADP-ribosyl)ation pathway is defective after treatment of H_2O_2 and MMS. Although PARP-1 becomes activated, the ADP-ribosylation of other cellular proteins is severely impaired. This finding suggests that the WRN/PARP-1 complex plays a role in the cellular response to oxidative stress and alkylating agents, implying a role for these proteins in the BER pathway.

Telomere Function

A reduced replicative potential compared with age-matched controls[105] is a hallmark of the WS cellular phenotypes. Since replicative senescence in human fibroblasts results from defects in telomere shortening and dysfunction,[17] premature senescence in WS cells may be consistent with telomere dysfunction. Expression of exogenous telomerase in WS fibroblasts confers extended cellular life span, lengthened telomeres, and replicative immortalization,[19,90,154] indicating that telomere effects predominantly trigger premature senescence in WS cells. Furthermore, expression of a telomerase transgene also partially reverses the hypersensitivity to 4-NQO in WS cells[46] although the underlying mechanism is not known.

The interactions of WRN with other proteins also indicate that WRN may participate in telomere metabolism. In fact, several proteins involved in DSB repair, including the Ku heterodimer and DNA-PKcs, as well as the Mre11-Rad50-Nbs1 complex that localize to telomeres[26,48,153] interact with WRN. Owing to this interaction, it has been proposed that WRN may function in a protein complex at telomeres. Indeed, Ku interacts with the critical telomere binding proteins TRF1 and TRF2,[49,121] and Nbs1 interacts with TRF2.[159] Both TRF1 and TRF2 regulate telomere length,[119] and defects in TRF2 induce telomere end fusions and either growth arrest or p53-mediated apoptosis.[55,56,144]

Furthermore, the WRN protein also interacts with TRF2 by coimmunoprecipitation and colocalization, and the purified proteins interact in vitro. The WRN helicase activity is stimulated by TRF2.[86] Although functions of these proteins in telomere maintenance are still not well understood, they may act in a cellular response to dysfunctional and/or damaged telomeres. Thus, WRN may act in cellular pathways that impact telomere integrity, and possibly respond to dysfunctional telomeres.

Structures at mammalian telomeres must be resolved to gain access to the terminal region of the telomere. Possible resolvases specific for telomere structures should recognize the loop-tail junction and/or interact with telomere binding proteins. As the WRN protein unwinds various DNA secondary structures including Holliday junctions,[75] G4-quartets that can be formed in the telomeric G-rich sequence,[75] as well as D-loop structures,[87] it may function in the resolution of telomeres during replication.

Cockayne's Syndrome

Cockayne syndrome (CS) is a rare inherited human genetic disease characterized by progressive multisystem degeneration, and categorized as a segmental progeria because of traits reminiscent of normal aging. This disorder is caused by mutations in the CSA[45] and CSB genes.[45,135] Affected individuals suffer from postnatal growth failure resulting in cachectic dwarfism, photosensitivity, skeletal abnormalities, mental retardation, progressive neurological degeneration, retinopathy, cataracts, and sensorineural hearing loss.[32,81] The majority of CS cases are caused by defects in the CSB protein. Cellular and biochemical studies suggest that the CSB protein is involved in several different processes, including general transcription, transcription-coupled NER (TCR), and BER of some types of oxidative damage;[25,63] (Fig. 4). Given the involvement of CSB in several different processes, it can be speculated that the participation of CSB in each pathway is somehow regulated in response to DNA damage, aging, development, and cellular differentiation.

Cellular features of Cockayne syndrome

DNA repair
        TCR defect after UV and cisplatin treatment
        Global BER defect of 8-oxoG and 8-oxoA
        Accumulation of oxidative lesions
        Sensitive to 4-NQO, NA-AAF

Transcription
        Basal transcription defect in vivo
        complex with RNA polymerase II
        Transcription elongation factor
        Lack or recovery of RNA synthesis after DNA damaging agent treatment
        UV, cisplatin, 4-NQO, NA-AAF

Chromatin remodeling

Apoptosis
        increased by UV treatment

Figure 4. Cellular features of Cockayne syndrome B. The CSB cellular responses to various DNA damaging agents and the affected cellular processes.

Biochemical Properties of the CS Proteins

The *CSA* and *CSB* genes have been cloned and their products are characterized biochemically. The *CSA* gene product is a 44 kDa protein that belongs to the "WD repeat" family of proteins.[45] Members of this protein family are structural and regulatory in function without any enzymatic activity. The *CSB* gene product is a 168 kDa protein[135] and is placed in the ERCC6 subfamily of the SNF2-like family, which contains seven conserved sequence motifs in DNA and RNA helicases domain.[92] Thus far, no helicase activity was found on a variety of different substrates.[21] Indeed, no member of the SNF2-like family has exhibited any helicase activity.[92] However, CSB exhibits DNA-dependent ATPase activity.[20,21,111] It is more stimulated by dsDNA than ssDNA and increased using a bubble or loop contained dsDNA, compared with closed dsDNA.[20] Even though CSB is involved in the NER process, damaged DNA, as a cofactor did not cause any significant change in the ATPase activity of CSB, indicating that CSB is not a damage-recognition factor on its own.

As a member of the SNF2-like family, which is all capable of chromatin remodeling, the ATPase activity is not unexpected with CSB. The investigation of CSB for chromatin remodeling revealed that CSB could remodel mononucleosomes and the chromatin structure on a chromatinized plasmid.[22] A recent study shows that CSB can be phosphorylated in vitro by CKII (casein kinase II) and in whole-extracts from human fibroblasts.[20] CSB is also phosphorylated in vivo; it can be dephosphorylated by UV irradiation. Interestingly, dephosphorylation of CSB increases its ATPase activity by 38%,[20] indicating phosphorylation-dependent regulation of CSB after DNA damage.

A Molecular Role of CSB in Ner Transcription-Coupled Repair

The precise molecular role of CSB is not clear at present. TCR by CSB is thought to remove transcription-blocking lesions possibly by recruitment of repair factors to the DNA lesion after removal of the stalled RNA polymerase.[128] In vitro, CSB exists in a quaternary complex with RNA pol II, DNA, and the RNA transcript; ATP hydrolysis is required to form this complex.[134] This quaternary complex recruits another molecular complex which includes the transcription factor IIH (TFIIH) core subunits, p62 and XPB.[133] TFIIH is a complex factor thought to promote local DNA unwinding during transcription initiation by RNA pol II and promoter escape, as well as in NER[30,31,107,108,145] Coimmunoprecipitation studies demonstrate that XPG and CSB proteins interact.[51] XPG interacts with multiple components of TFIIH[52] and it is an endonuclease that plays a role in NER. In vitro, CSB interacts with CSA and with the NER damage recognition factor XPA.[45,110] These protein interactions support the hypothesis that CSB participates directly in TCR.

It is also possible that CSB indirectly stimulates TCR by facilitating transcription. Members of the SWI/SNF family are involved in regulating transcription, chromatin remodeling, and DNA repair, including such actions as disruption of protein-protein and protein-DNA interactions (reviewed for ref.[92]). The *CSB* gene product could mediate change in chromatin structure in an ATP-dependent manner and modify the interaction between stalled RNA pol II and DNA. Thus, this activity allows repair of a polymerase-blocking lesion and/or resumption of elongation.[22] In fact, it is still a matter of debate whether CS is due to a primary defect in transcription or DNA repair.[141,142] The basal transcription level is lower in human CS-B lymphoblastoid cells and fibroblasts than in normal cells, even without exposure to DNA damaging agents.[3] This transcription defect is complemented by normal cell extracts in vitro or by the wildtype *CSB* gene in vivo. In a reconstituted system, purified CSB protein enhances the rate of transcription by RNA pol II,[110] suggesting that CSB may indirectly stimulate TCR by facilitating the process of transcription.[111] Thus, CSB may be a transcription elongation factor and repair-coupling factor acting at the site of RNA pol II-blocking lesions, and the CS phenotype may arise from deficiencies in both transcription and DNA repair. Distinct functional domains of the protein may mediate the biological actions of CSB in these different pathways.

CS cell lines and primary cells are sensitive to UV light,[135,150] 4-NQO,[78,138,150] NA-AAF (N-acetoxy-2-acetylaminofluorene),[124,143] and ionizing radiation.[27] Cyclobutane pyrimidine dimer (CPD) photoproducts are mainly removed by TCR, whereas 6-4 photoproducts, the DNA lesions inflicted by 4-NQO[120] and NA-AAF,[132] are predominantly removed by global genome repair (GGR).

A characteristic cellular phenotype of CSB cells is the deficient recovery of RNA synthesis after exposure to UV.[135,140] RNA synthesis in normal cells is transiently inhibited following UV exposure, but unlike CSB cells, they recover well. The defect in RNA synthesis recovery after UV irradiation is thought to reflect a defect in TCR. Also, primary CS cells are deficient in recovery of RNA synthesis after treatment with paraoxon/NA-AAF or 4-NQO compared with WT cells.[9,143] This indeed is a surprising finding because these lesions are considered to be substrates for GGR. Thus, it has been proposed that CSB acts as a transcription-repair-uncoupling factor, possibly by helping TFIIH switch from a repair to a transcription mode.

A Molecular Role of CSB in BER

As described above, it is well established that the *CSB* phenotype involves a defect in TCR of UV-induced DNA damage. However, evidence also suggests that TCR of oxidative damage may be affected by CSB.[62] Hydrogen peroxide and ionizing radiation induce a large variety of DNA lesions including oxidatively damaged bases, single- and double-strand DNA breaks. It has been shown that primary CS-B cells are slightly more sensitive to γ-irradiation; and in addition, they are defective in strand-specific repair of ionizing radiation and hydrogen peroxide-induced thymine glycol.[25,63] Previously, this laboratory showed that the ability of CS-B cell extracts to carry out incision of 8-oxoguanine lesions in vitro is 50% of normal; in contrast, incision of uracil and thymine glycol were normal.[29] 8-Oxoguanine, a base modification caused by oxidative stress from environmental agents and endogenous metabolic processes, is repaired mainly via by BER.[28,67] Further recent studies from this laboratory show that the CSB protein is involved in general genome BER. We have established stably transfected human cell lines in which functional domains of the *CSB* gene have been mutated.[138] When certain helicase motifs are disrupted, there is a cellular defect in the incision of 8-oxoG containing DNA in cell extracts[138] and 8-oxoA[137] after treatment with γ radiation. In addition, cells from CS patients accumulated significant amounts of these lesions and the whole extracts of fibroblasts from the patients were repair deficient.[136] Together, a deficiency in cellular repair of oxidative DNA damage might contribute to developmental defects in CS patients.

Understanding the exact role of CSB is quite challenging. The multisystem character of CS and the complexity of the genotype-phenotype relationship give clues to the pathways that involve CSB. In support of this, CSB interacts physically, functionally with proteins in different processes. We are convinced that this important protein participates in not only transcription-coupled repair, but also BER, and that it also plays a role in transcription. Characterization of the biological functions of different domains of the CSB protein would shed lights on the understanding of the CSB phenotype.

Conclusions

It is clear from the foregoing account that mutations in genes that encode helicase domain-containing proteins lead to the onset of human disorders that display accelerated aging phenotypes. Genetic mutations giving rise to the progeroid symptoms in these patients partially or wholly inactivate proteins that sense or repair the DNA damage. This observation suggests that failure to maintain genome integrity may be a primary cause for the aging phenotypes.

Although the precise role of the WRN in DNA repair remains to be clarified, recent data indicate that WRN has multiple roles in genome maintenance. Cells from WS patients with defects in the genomic maintenance pathways often show both the apoptotic and senescence responses to different types of DNA damage. CSB functions as a general RNA polymerase II transcription factor and an integral component of NER and BER machineries. Since a general decline in transcription occurs in a variety of organs and tissues with increasing age, transcription deregulation manifested in CSB due to unrepaired DNA damage is likely to be responsible for some of the aging related symptoms. CS patients in general do not live long and even during their short life span, they exhibit accelerated aging features suggestive of a vital function for CSB protein for both longevity and overall genome maintenance during the lifetime of humans.

To understand the basis of accelerated aging and normal aging, it is important to consider the molecular and cellular changes that can result from defective genome maintenance. Cells from humans with defective DNA repair machinery may face the consequences of unrepaired DNA damage in the form of apoptosis or cellular senescence. Thus, increased loss of functional cells in organs and tissues may contribute to aging. It is not clear whether the same scenario also occurs during the normal aging process due to age dependent decline in DNA repair efficiency. Understanding the precise biological functions of WRN and CSB genes and the knowledge of primary and secondary effects of mutations in these genes will provide insights into the molecular mechanism(s) of premature aging and normal aging processes.

Acknowledgements

This work was supported in part by the Basic Research Program Grant (R01-2003-000-00361) from the Korea Science and Engineering Foundation to B. Ahn.

References

1. Adelfalk C, Kontou M, Hirsch-Kauffmann M et al. Physical and functional interaction of the Werner syndrome protein with poly-ADP ribosyl transferase. FEBS Lett 2003; 554:55-58.
2. Balajee AS, Machwe A, May A et al. The Werner syndrome protein is involved in RNA polymerase II transcription. Mol Biol Cell 1999; 10:2655-2668.
3. Balajee AS, May A, Dianov GL et al. Reduced RNA polymerase II transcription in intact and permeabilized Cockayne syndrome group B cells. Proc Natl Acad Sci USA 1997; 94:4306-4311.
4. Bambara RA, Murante RS, Henricksen LA. Enzymes and reactions at the eukaryotic DNA replication fork. J Biol Chem 1997; 272:4647-4650.
5. Baynton K, Otterlei M, Bjoras M et al. WRN interacts physically and functionally with the recombination mediator protein RAD52. J Biol Chem 2003; 278:36476-36486.
6. Blander G, Zalle N, Daniely Y et al. DNA damage-induced translocation of the Werner helicase is regulated by acetylation. J Biol Chem 2002; 277:50934-50940.
7. Blander G, Zalle N, Leal JF et al. The Werner syndrome protein contributes to induction of p53 by DNA damage. FASEB J 2000; 14:2138-2140.
8. Bohr VA, Souza PN, Nyaga SG et al. DNA repair and mutagenesis in Werner syndrome. Environ Mol Mutagen 2001; 38:227-234.
9. Brosh Jr RM, Balajee AS, Selzer RR et al. The ATPase domain but not the acidic region of Cockayne syndrome group B gene product is essential for DNA repair. Mol Biol Cell 1999a; 10:3583-3594.
10. Brosh Jr RM, Driscoll HC, Dianov GL et al. Biochemical characterization of the WRN-FEN-1 functional interaction. Biochemistry 2002; 41:12204-12216.
11. Brosh Jr RM, Karmakar P, Sommers JA et al. p53 Modulates the exonuclease activity of Werner syndrome protein. J Biol Chem 2001a; 276:35093-35102.
12. Brosh Jr RM, Li JL, Kenny MK et al. Replication protein A physically interacts with the Bloom's syndrome protein and stimulates its helicase activity. J Biol Chem 2000; 275:23500-23508.
13. Brosh Jr RM, Majumdar A, Desai S et al. Unwinding of a DNA triple helix by the Werner and Bloom syndrome helicases. J Biol Chem 2001b; 276:3024-3030.

14. Brosh Jr RM, Orren DK, Nehlin JO et al. Functional and physical interaction between WRN helicase and human replication protein A. J Biol Chem 1999b; 274:18341-18350.
15. Brosh Jr RM, von Kobbe C, Sommers JA et al. Werner syndrome protein interacts with human flap endonuclease 1 and stimulates its cleavage activity. EMBO J 2001c; 20:5791-5801.
16. Burkle A. Poly(APD-ribosyl)ation, a DNA damage-driven protein modification and regulator of genomic instability. Cancer Lett 2001; 163:1-5.
17. Campisi J, Kim SH, Lim CS et al. Cellular senescence, cancer and aging: The telomere connection. Exp Gerontol 2001; 36:1619-1637.
18. Cheng WH, von Kobbe C, Opresko PL et al. Linkage between Werner syndrome protein and the Mre11 complex via Nbs1. J Biol Chem 2004; 279:21169-21176.
19. Choi D, Whittier PS, Oshima J et al. Telomerase expression prevents replicative senescence but does not fully reset mRNA expression patterns in Werner syndrome cell strains. FASEB J 2001; 15:1014-1020.
20. Christiansen M, Stevnsner T, Modin C et al. Functional consequences of mutations in the conserved SF2 motifs and post-translational phosphorylation of the CSB protein. Nucleic Acids Res 2003; 31:963-973.
21. Citterio E, Rademakers S, van der Horst GT et al. Biochemical and biological characterization of wild-type and ATPase-deficient Cockayne syndrome B repair protein. J Biol Chem 1998; 273:11844-11851.
22. Citterio E, Van DBV, Schnitzler G et al. ATP-dependent chromatin remodeling by the Cockayne syndrome B DNA repair-transcription-coupling factor. Mol Cell Biol 2000; 20:7643-7653.
23. Constantinou A, Tarsounas M, Karow JK et al. Werner's syndrome protein (WRN) migrates Holliday junctions and colocalizes with RPA upon replication arrest. EMBO Rep 2000; 1:80-84.
24. Cooper MP, Machwe A, Orren DK et al. Ku complex interacts with and stimulates the Werner protein. Genes Dev 2000; 14:907-912.
25. Cooper PK, Nouspikel T, Clarkson SG et al. Defective transcription-coupled repair of oxidative base damage in Cockayne syndrome patients from XP group G. Science 1997; 275:990-993.
26. d'Adda dF, Hande MP, Tong WM et al. Effects of DNA nonhomologous end-joining factors on telomere length and chromosomal stability in mammalian cells. Curr Biol 2001; 11:1192-1196.
27. de Waard H, de Wit J, Gorgels TG et al. Cell type-specific hypersensitivity to oxidative damage in CSB and XPA mice. DNA Repair (Amst) 2003; 2:13-25.
28. Dianov G, Bischoff C, Piotrowski J et al. Repair pathways for processing of 8-oxoguanine in DNA by mammalian cell extracts. J Biol Chem 1998; 273:33811-33816.
29. Dianov G, Bischoff C, Sunesen M et al. Repair of 8-oxoguanine in DNA is deficient in Cockayne syndrome group B cells. Nucleic Acids Res 1999; 27:1365-1368.
30. Drapkin R, Le Roy G, Cho H et al. Human cyclin-dependent kinase-activating kinase exists in three distinct complexes. Proc Natl Acad Sci USA 1996; 93:6488-6493.
31. Feaver WJ, Svejstrup JQ, Bardwell L et al. Dual roles of a multiprotein complex from S. cerevisiae in transcription and DNA repair. Cell 1993; 75:1379-1387.
32. Friedberg EC. Cockayne syndrome—a primary defect in DNA repair, transcription, both or neither? Bioessays 1996; 18:731-738.
33. Fry M, Loeb LA. Human werner syndrome DNA helicase unwinds tetrahelical structures of the fragile X syndrome repeat sequence d(CGG)n. J Biol Chem 1999; 274:12797-12802.
34. Fujiwara Y, Higashikawa T, Tatsumi M. A retarded rate of DNA replication and normal level of DNA repair in Werner's syndrome fibroblasts in culture. J Cell Physiol 1977; 92:365-374.
35. Fukuchi K, Martin GM, Monnat Jr RJ. Mutator phenotype of Werner syndrome is characterized by extensive deletions. Proc Natl Acad Sci USA 1989; 86:5893-5897.
36. Gebhart E, Bauer R, Raub U et al. Spontaneous and induced chromosomal instability in Werner syndrome. Hum Genet 1988; 80:135-139.
37. Gharibyan V, Youssoufian H. Localization of the Bloom syndrome helicase to punctate nuclear structures and the nuclear matrix and regulation during the cell cycle: Comparison with the Werner's syndrome helicase. Mol Carcinog 1999; 26:261-273.
38. Gorbalenya AE, Koonin EV, Donchenko AP et al. Two related superfamilies of putative helicases involved in replication, recombination, repair and expression of DNA and RNA genomes. Nucleic Acids Res 1989; 17:4713-4730.

39. Gray MD, Shen JC, Kamath-Loeb AS et al. The Werner syndrome protein is a DNA helicase. Nat Genet 1997; 17:100-103.

40. Gray MD, Wang L, Youssoufian H et al. Werner helicase is localized to transcriptionally active nucleoli of cycling cells. Exp Cell Res 1998; 242:487-494.

41. Haber JE. DNA recombination: The replication connection. Trends Biochem Sci 1999; 24:271-275.

42. Hanaoka F, Yamada M, Takeuchi F et al. Autoradiographic studies of DNA replication in Werner's syndrome cells. Adv Exp Med Biol 1985; 190:439-457.

43. Harmon FG, Kowalczykowski SC. RecQ helicase, in concert with RecA and SSB proteins, initiates and disrupts DNA recombination. Genes Dev 1998; 12:1134-1144.

44. Harrigan JA, Opresko PL, von Kobbe C et al. The Werner syndrome protein stimulates DNA polymerase beta strand displacement synthesis via its helicase activity. J Biol Chem 2003; 278:22686-22695.

45. Henning KA, Li L, Iyer N et al. The Cockayne syndrome group A gene encodes a WD repeat protein that interacts with CSB protein and a subunit of RNA polymerase II TFIIH. Cell 1995; 82:555-564.

46. Hisama FM, Chen YH, Meyn MS et al. WRN or telomerase constructs reverse 4-nitroquinoline 1-oxide sensitivity in transformed Werner syndrome fibroblasts. Cancer Res 2000; 60:2372-2376.

47. Hoehn H, Bryant EM, Au K et al. Variegated translocation mosaicism in human skin fibroblast cultures. Cytogenet Cell Genet 1975; 15:282-298.

48. Hsu HL, Gilley D, Blackburn EH et al. Ku is associated with the telomere in mammals. Proc Natl Acad Sci USA 1999; 96:12454-12458.

49. Hsu HL, Gilley D, Galande SA et al. Ku acts in a unique way at the mammalian telomere to prevent end joining. Genes Dev 2000; 14:2807-2812.

50. Huang S, Li B, Gray MD et al. The premature ageing syndrome protein, WRN, is a 3'—>5' exonuclease. Nat Genet 1998; 20:114-116.

51. Iyer N, Reagan MS, Wu KJ et al. Interactions involving the human RNA polymerase II transcription/nucleotide excision repair complex TFIIH, the nucleotide excision repair protein XPG, and Cockayne syndrome group B (CSB) protein. Biochemistry 1996b; 35:2157-2167.

52. Iyer N, Reagan MS, Wu KJ et al. Interactions involving the human RNA polymerase II transcription/nucleotide excision repair complex TFIIH, the nucleotide excision repair protein XPG, and Cockayne syndrome group B (CSB) protein. Biochemistry 1996a; 35:2157-2167.

53. Kamath-Loeb AS, Shen JC, Loeb LA et al. Werner syndrome protein. II. Characterization of the integral 3' —> 5' DNA exonuclease. J Biol Chem 1998; 273:34145-34150.

54. Kao HI, Henricksen LA, LiuY et al. Cleavage specificity of Saccharomyces cerevisiae flap endonuclease 1 suggests a double-flap structure as the cellular substrate. J Biol Chem 2002; 277:14379-14389.

55. Karlseder J, Broccoli D, Dai Y et al. p53- and ATM-dependent apoptosis induced by telomeres lacking TRF2. Science 1999; 283:1321-1325.

56. Karlseder J, Smogorzewska A, de Lange T. Senescence induced by altered telomere state, not telomere loss. Science 2002; 295:2446-2449.

57. Khakhar RR, Cobb JA, Bjergbaek L et al. RecQ helicases: Multiple roles in genome maintenance. Trends Cell Biol 2003; 13:493-501.

58. Kruk PA, Rampino NJ, Bohr VA. DNA damage and repair in telomeres: Relation to aging. Proc Natl Acad Sci USA 1995; 92:258-262.

59. Kusano K, Berres ME, Engels WR. Evolution of the RECQ family of helicases: A drosophila homolog, Dmblm, is similar to the human bloom syndrome gene. Genetics 1999; 151:1027-1039.

60. Kyng KJ, May A, Kolvraa S et al. Gene expression profiling in Werner syndrome closely resembles that of normal aging. Proc Natl Acad Sci USA 2003; 100:12259-12264.

61. Laine JP, Opresko PL, Indig FE et al. Werner protein stimulates topoisomerase I DNA relaxation activity. Cancer Res 2003; 63:7136-7146.

62. Le Page F, Kwoh EE, Avrutskaya A et al. Transcription-coupled repair of 8-oxoguanine: Requirement for XPG, TFIIH, and CSB and implications for Cockayne syndrome. Cell 2000; 101:159-171.

63. Leadon SA, Cooper PK. Preferential repair of ionizing radiation-induced damage in the transcribed strand of an active human gene is defective in Cockayne syndrome. Proc Natl Acad Sci USA 1993; 90:10499-10503.

64. Lebel M, Leder P. A deletion within the murine Werner syndrome helicase induces sensitivity to inhibitors of topoisomerase and loss of cellular proliferative capacity. Proc Natl Acad Sci USA 1998; 95:13097-13102.

65. Lebel M, Spillare EA, Harris CC et al. The Werner syndrome gene product copurifies with the DNA replication complex and interacts with PCNA and topoisomerase I. J Biol Chem 1999; 274:37795-37799.

66. Lieber MR, Ma Y, Pannicke U et al. Mechanism and regulation of human nonhomologous DNA end-joining. Nat Rev Mol Cell Biol 2003; 4:712-720.

67. Lindahl T. Instability and decay of the primary structure of DNA. Nature 1993; 362:709-715.

68. Lu J, Mullen JR, Brill SJ et al. Human homologues of yeast helicase. Nature 1996; 383:678-679.

69. Martin GM. Genetic syndromes in man with potential relevance to the pathobiology of aging. Birth Defects Orig Artic Ser 1978; 14:5-39.

70. Martin GM, Sprague CA, Epstein CJ. Replicative life-span of cultivated human cells. Effects of donor's age, tissue, and genotype. Lab Invest 1970; 23:86-92.

71. Matsumoto T, Shimamoto A, Goto M et al. Impaired nuclear localization of defective DNA helicases in Werner's syndrome. Nat Genet 1997; 16:335-336.

72. McGlynn P, Lloyd RG. Recombinational repair and restart of damaged replication forks. Nat Rev Mol Cell Biol 2002; 3:859-870.

73. McVey M, Kaeberlein M, Tissenbaum HA et al. The short life span of Saccharomyces cerevisiae sgs1 and srs2 mutants is a composite of normal aging processes and mitotic arrest due to defective recombination. Genetics 2001; 157:1531-1542.

74. Melaragno MI, Pagni D, Smith MA. Cytogenetic aspects of Werner's syndrome lymphocyte cultures. Mech Ageing Dev 1995; 78:117-122.

75. Mohaghegh P, Karow JK, Brosh JR et al. The Bloom's and Werner's syndrome proteins are DNA structure specific helicases. Nucleic Acids Res 2001; 29:2843-2849.

76. Morozov V, Mushegian AR, Koonin EV et al. A putative nucleic acid-binding domain in Bloom's and Werner's syndrome helicases. Trends Biochem Sci 1997; 22:417-418.

77. Moser MJ, Holley WR, Chatterjee A et al. The proofreading domain of Escherichia coli DNA polymerase I and other DNA and/or RNA exonuclease domains. Nucleic Acids Res 1997; 25:5110-5118.

78. Muftuoglu M, Selzer R, Tuo J et al. Phenotypic consequences of mutations in the conserved motifs of the putative helicase domain of the human Cockayne syndrome group B gene. Gene 2002; 283:27-40.

79. Mushegian AR, Bassett Jr DE, Boguski MS et al. Positionally cloned human disease genes: Patterns of evolutionary conservation and functional motifs. Proc Natl Acad Sci USA 1997; 94:5831-5836.

80. Myung K, Datta A, Chen C et al. SGS1, the Saccharomyces cerevisiae homologue of BLM and WRN, suppresses genome instability and homeologous recombination. Nat Genet 2001; 27:113-116.

81. Nance MA, Berry SA. Cockayne syndrome: Review of 140 cases. Am J Med Genet 1992; 42:68-84.

82. Ogburn CE, Oshima J, Poot M et al. An apoptosis-inducing genotoxin differentiates heterozygotic carriers for Werner helicase mutations from wild-type and homozygous mutants. Hum Genet 1997; 101:121-125.

83. Okada M, Goto M, Furuichi Y et al. Differential effects of cytotoxic drugs on mortal and immortalized B-lymphoblastoid cell lines from normal and Werner's syndrome patients. Biol Pharm Bull 1998; 21:235-239.

84. Opresko PL, Cheng WH, von Kobbe C et al. Werner syndrome and the function of the Werner protein; what they can teach us about the molecular aging process. Carcinogenesis 2003; 24:791-802.

85. Opresko PL, Laine JP, Brosh Jr RM et al. Coordinate action of the helicase and 3' to 5' exonuclease of Werner syndrome protein. J Biol Chem 2001; 276:44677-44687.

86. Opresko PL, von Kobbe C, Laine JP et al. Telomerebinding protein TRF2 binds to and stimulates the Werner and Bloom syndrome helicases. J Biol Chem 2002; 277:41110-41119.

86a. Opresko PL, Oyyellei M, Groakjars J et al. The Werner syndrome helicase and exonuclease cooperate to resolve telomeric D-loops in a manner regulated by TRF1 and TRF2. Mol Cell 2004; 14:763-774.

87. Orren DK, Theodore S, Machwe A. The Werner syndrome helicase/exonuclease (WRN) disrupts and degrades D-loops in vitro. Biochemistry 2002; 41:13483-13488.

88. Oshima J. The Werner syndrome protein: An update. Bioessays 2000; 22:894-901.

89. Oshima J, Huang S, Pae C et al. Lack of WRN results in extensive deletion at nonhomologous joining ends. Cancer Res 2002; 62:547-551.

90. Ouellette MM, McDaniel LD, Wright WE et al. The establishment of telomerase-immortalized cell lines representing human chromosome instability syndromes. Hum Mol Genet 2000; 9:403-411.

91. Partridge JJ, Lopreiato Jr JO, Latterich M et al. DNA damage modulates nucleolar interaction of the Werner protein with the AAA ATPase p97/VCP. Mol Biol Cell 2003; 14:4221-4229.

92. Pazin MJ, Kadonaga JT. SWI2/SNF2 and related proteins: ATP-driven motors that disrupt protein-DNA interactions? Cell 1997; 88:737-740.

93. Pichierri P, Franchitto A, Mosesso P et al. Werner's syndrome protein is required for correct recovery after replication arrest and DNA damage induced in S-phase of cell cycle. Mol Biol Cell 2001; 12:2412-2421.

94. Pichierri P, Franchitto A, Mosesso P et al. Werner's syndrome lymphoblastoid cells are hypersensitive to topoisomerase II inhibitors in the G2 phase of the cell cycle. Mutat Res 2000; 459:123-133.

95. Poot M, Gollahon KA, Emond MJ et al. Werner syndrome diploid fibroblasts are sensitive to 4-nitroquinoline-N-oxide and 8-methoxypsoralen: Implications for the disease phenotype. FASEB J 2002; 16:757-758.

96. Poot M, Gollahon KA, Rabinovitch PS. Werner syndrome lymphoblastoid cells are sensitive to camptothecin-induced apoptosis in S-phase. Hum Genet 1999; 104:10-14.

97. Poot M, Hoehn H, Runger TM et al. Impaired S-phase transit of Werner syndrome cells expressed in lymphoblastoid cell lines. Exp Cell Res 1992; 202:267-273.

98. Poot M, Yom JS, Whang SH et al. Werner syndrome cells are sensitive to DNA cross-linking drugs. FASEB J 2001; 15:1224-1226.

99. Prince PR, Emond MJ, Monnat Jr RJ. Loss of Werner syndrome protein function promotes aberrant mitotic recombination. Genes Dev 2001; 15:933-938.

100. Rothstein R, Michel B, Gangloff S. Replication fork pausing and recombination or "gimme a break". Genes Dev 2000; 14:1-10.

101. Saffi J, Pereira VR, Henriques JA. Importance of the Sgs1 helicase activity in DNA repair of Saccharomyces cerevisiae. Curr Genet 2000; 37:75-78.

102. Saintigny Y, Makienko K, Swanson C et al. Homologous recombination resolution defect in werner syndrome. Mol Cell Biol 2002; 22:6971-6978.

103. Sakamoto S, Nishikawa K, Heo SJ et al. Werner helicase relocates into nuclear foci in response to DNA damaging agents and colocalizes with RPA and Rad51. Genes Cells 2001; 6:421-430.

104. Salk D. Werner's syndrome: A review of recent research with an analysis of connective tissue metabolism, growth control of cultured cells, and chromosomal aberrations. Hum Genet 1982; 62:1-5.

105. Salk D, Au K, Hoehn H et al. Cytogenetics of Werner's syndrome cultured skin fibroblasts: Variegated translocation mosaicism. Cytogenet Cell Genet 1981; 30:92-107.

106. Salk D, Bryant E, Hoehn H et al. Growth characteristics of Werner syndrome cells in vitro. Adv Exp Med Biol 1985; 190:305-311.

107. Schaeffer L, Moncollin V, Roy R et al. The ERCC2/DNA repair protein is associated with the class II BTF2/TFIIH transcription factor. EMBO J 1994; 13:2388-2392.

108. Schaeffer L, Roy R, Humbert S et al. DNA repair helicase: A component of BTF2 (TFIIH) basic transcription factor. Science 1993; 260:58-63.

109. Schulz VP, Zakian VA, Ogburn CE et al. Accelerated loss of telomeric repeats may not explain accelerated replicative decline of Werner syndrome cells. Hum Genet 1996; 97:750-754.

110. Selby CP, Drapkin R, Reinberg D et al. RNA polymerase II stalled at a thymine dimer: Footprint and effect on excision repair. Nucleic Acids Res 1997; 25:787-793.

111. Selby CP, Sancar A. Human transcription-repair coupling factor CSB/ERCC6 is a DNA-stimulated ATPase but is not a helicase and does not disrupt the ternary transcription complex of stalled RNA polymerase II. J Biol Chem 1997; 272:1885-1890.

112. Sharma S, Otterlei M, Sommers JA et al. WRN Helicase and FEN-1 form a complex upon replication arrest and together process branchmigrating DNA structures associated with the replication fork. Mol Biol Cell 2004; 15:734-750.

113. Shen JC, Gray MD, Oshima J et al. Werner syndrome protein. I. DNA helicase and dna exonuclease reside on the same polypeptide. J Biol Chem 1998a; 273:34139-34144.
114. Shen JC, Gray MD, Oshima J et al. Characterization of Werner syndrome protein DNA helicase activity: Directionality, substrate dependence and stimulation by replication protein A. Nucleic Acids Res 1998b; 26:2879-2885.
115. Shen JC, Lao Y, Kamath-Loeb A et al. The N-terminal domain of the large subunit of human replication protein A binds to Werner syndrome protein and stimulates helicase activity. Mech Ageing Dev 2003; 124:921-930.
116. Shen JC, Loeb LA. Werner syndrome exonuclease catalyzes structure dependent degradation of DNA. Nucleic Acids Res 2000; 28:3260-3268.
117. Shiratori M, Suzuki T, Itoh C et al. WRN helicase accelerates the transcription of ribosomal RNA as a component of an RNA polymerase I-associated complex. Oncogene 2002; 21:2447-2454.
118. Sinclair DA, Mills K, Guarente L. Accelerated aging and nucleolar fragmentation in yeast sgs1 mutants. Science 1997; 277:1313-1316.
119. Smogorzewska A, van Steensel B, Bianchi A et al. Control of human telomere length by TRF1 and TRF2. Mol Cell Biol 2000; 20:1659-1668.
120. Snyderwine EG, Bohr VA. Gene- and strand-specific damage and repair in Chinese hamster ovary cells treated with 4-nitroquinoline 1-oxide. Cancer Res 1992; 52:4183-4189.
121. Song K, Jung D, Jung Y et al. Interaction of human Ku70 with TRF2. FEBS Lett 2000; 481:81-85.
122. Spillare EA, Robles AI, Wang XW et al. p53-mediated apoptosis is attenuated in Werner syndrome cells. Genes Dev 1999; 13:1355-1360.
123. Stefanini M, Scappaticci S, Lagomarsini P et al. Chromosome instability in lymphocytes from a patient with Werner's syndrome is not associated with DNA repair defects. Mutat Res 1989; 219:179-185.
124. Sunesen M, Selzer RR, Brosh Jr RM et al. Molecular characterization of an acidic region deletion mutant of Cockayne syndrome group B protein. Nucleic Acids Res 2000; 28:3151-3159.
125. Suzuki N, Shimamoto A, Imamura O et al. DNA helicase activity in Werner's syndrome gene product synthesized in a baculovirus system. Nucleic Acids Res 1997; 25:2973-2978.
126. Suzuki N, Shiratori M, Goto M et al. Werner syndrome helicase contains a 5'—>3' exonuclease activity that digests DNA and RNA strands in DNA/DNA and RNA/DNA duplexes dependent on unwinding. Nucleic Acids Res 1999; 27:2361-2368.
127. Suzuki T, Shiratori M, Furuichi Y et al. Diverged nuclear localization of Werner helicase in human and mouse cells. Oncogene 2001; 20:2551-2558.
128. Svejstrup JQ. Mechanisms of transcription-coupled DNA repair. Nat Rev Mol Cell Biol 2002; 3:21-29.
129. Szekely AM, Chen YH, Zhang C et al. Werner protein recruits DNA polymerase delta to the nucleolus. Proc Natl Acad Sci USA 2000; 97:11365-11370.
130. Tahara H, Tokutake Y, Maeda S et al. Abnormal telomere dynamics of B-lymphoblastoid cell strains from Werner's syndrome patients transformed by Epstein-Barr virus. Oncogene 1997; 15:1911-1920.
131. Takeuchi F, Hanaoka F, Goto M et al. Altered frequency of initiation sites of DNA replication in Werner's syndrome cells. Hum Genet 1982; 60:365-368.
132. Tang MS, Bohr VA, Zhang XS et al. Quantification of aminofluorene adduct formation and repair in defined DNA sequences in mammalian cells using the UvrABC nuclease. J Biol Chem 1989; 264:14455-14462.
133. Tantin D. RNA polymerase II elongation complexes containing the Cockayne syndrome group B protein interact with a molecular complex containing the transcription factor IIH components xeroderma pigmentosum B and p62. J Biol Chem 1998; 273:27794-27799.
134. Tantin D, Kansal A, Carey M. Recruitment of the putative transcription-repair coupling factor CSB/ERCC6 to RNA polymerase II elongation complexes. Mol Cell Biol 1997; 17:6803-6814.
135. Troelstra C, van Gool A, de Wit J et al. ERCC6, a member of a subfamily of putative helicases, is involved in Cockayne's syndrome and preferential repair of active genes. Cell 1992; 71:939-953.
136. Tuo J, Jaruga P, Rodriguez H et al. Primary fibroblasts of Cockayne syndrome patients are defective in cellular repair of 8-hydroxyguanine and 8-hydroxyadenine resulting from oxidative stress. FASEB J 2003; 17:668-674.

137. Tuo J, Jaruga P, Rodriguez H et al. The cockayne syndrome group B gene product is involved in cellular repair of 8-hydroxyadenine in DNA. J Biol Chem 2002; 277:30832-30837.

138. Tuo J, Muftuoglu M, Chen C et al. The Cockayne Syndrome group B gene product is involved in general genome base excision repair of 8-hydroxyguanine in DNA. J Biol Chem 2001; 276:45772-45779.

139. Umezu K, Nakayama K, Nakayama H. Escherichia coli RecQ protein is a DNA helicase. Proc Natl Acad Sci USA 1990; 87:5363-5367.

140. van der Horst GT, Meira L, Gorgels TG et al. UVB radiation-induced cancer predisposition in Cockayne syndrome group A (CSA) mutant mice. DNA Repair (Amst) 2002; 1:143-157.

141. van Gool AJ, Citterio E, Rademakers S et al. The Cockayne syndrome B protein, involved in transcription-coupled DNA repair, resides in an RNA polymerase II-containing complex. EMBO J 1997; 16:5955-5965.

142. van Gool AJ, Verhage R, Swagemakers SM et al. RAD26, the functional S. cerevisiae homolog of the Cockayne syndrome B gene ERCC6. EMBO J 1994; 13:5361-5369.

143. van Oosterwijk MF, Filon R, de Groot AJ et al. Lack of transcription-coupled repair of acetylaminofluorene DNA adducts in human fibroblasts contrasts their efficient inhibition of transcription. J Biol Chem 1998; 273:13599-13604.

144. van Steensel B, Smogorzewska A, de Lange T. TRF2 protects human telomeres from end-to-end fusions. Cell 1998; 92:401-413.

145. van Vuuren AJ, Vermeulen W, Ma L et al. Correction of xeroderma pigmentosum repair defect by basal transcription factor BTF2 (TFIIH). EMBO J 1994; 13:1645-1653.

146. von Kobbe C, Bohr VA. A nucleolar targeting sequence in the Werner syndrome protein resides within residues 949-1092. J Cell Sci 2002; 115:3901-3907.

147. von Kobbe C, Harrigan JA, May A et al. Central role for the Werner syndrome protein/ poly(ADP-ribose) polymerase 1 complex in the poly(ADP-ribosyl)ation pathway after DNA damage. Mol Cell Biol 2003a; 23:8601-8613.

148. von Kobbe C, Thoma NH, Czyzewski BK et al. Werner syndrome protein contains three structure specific DNA binding domains. J Biol Chem 2003b; 278:52997-53006.

149. von Zglinicki T, Burkle A, Kirkwood TB. Stress, DNA damage and ageing — an integrative approach. Exp Gerontol 2001; 36:1049-1062.

150. Wade MH, Chu EH. Effects of DNA damaging agents on cultured fibroblasts derived from patients with Cockayne syndrome. Mutat Res 1979; 59:49-60.

151. Watt PM, Hickson ID, Borts RH et al. SGS1, a homologue of the Bloom's and Werner's syndrome genes, is required for maintenance of genome stability in Saccharomyces cerevisiae. Genetics 1996; 144:935-945.

152. Werner O. On cataract associated in conjuction with scleroderma. Kiel University 1904, (ref. type: Thesis/Dissertation).

153. Wu G, Lee WH, Chen PL. NBS1 and TRF1 colocalize at promyelocytic leukemia bodies during late S/G2 phases in immortalized telomerase-negative cells. Implication of NBS1 in alternative lengthening of telomeres. J Biol Chem 2000; 275:30618-30622.

154. Wyllie FS, Jones CJ, Skinner JW et al. Telomerase prevents the accelerated cell ageing of Werner syndrome fibroblasts. Nat Genet 2000; 24:16-17.

155. Xie Y, Liu Y, Argueso JL et al. Identification of rad27 mutations that confer differential defects in mutation avoidance, repeat tract instability, and flap cleavage. Mol Cell Biol 2001; 21:4889-4899.

156. Yamagata K, Kato J, Shimamoto A et al. Bloom's and Werner's syndrome genes suppress hyperrecombination in yeast sgs1 mutant: Implication for genomic instability in human diseases. Proc Natl Acad Sci USA 1998; 95:8733-8738.

157. Yannone SM, Roy S, Chan DW et al. Werner syndrome protein is regulated and phosphorylated by DNA-dependent protein kinase. J Biol Chem 2001; 276:38242-38248.

158. Yu CE, Oshima J, Fu YH et al. Positional cloning of the Werner's syndrome gene. Science 1996; 272:258-262.

159. Zhu XD, Kuster B, Mann M et al. Cell-cycle-regulated association of RAD50/MRE11/NBS1 with TRF2 and human telomeres. Nat Genet 2000; 25:347-352.

CHAPTER 2

DNA Repair Aspects for RecQ Helicase Disorders

Takehisa Matsumoto

Abstract

RecQ family DNA helicases are defined by amino acid sequence similarities to *Escherichia coli* RecQ which has been known to act in homologous recombination and to suppress illegitimate recombination, particularly during the repair of DNA double strand breaks. Five *RecQ* family genes have been identified in humans, and three (*BLM*, *WRN*, and *RECQL4*) have been identified as defective in the human genetic disorders; Bloom syndrome, Werner syndrome, and a subset of Rothmund-Thomson syndrome. Despite strong homology in the helicase domains, human *RecQ* family genes differ markedly outside these domains. Indeed, each syndrome presents different phenotypes. however, all are characterized by an increased predisposition to cancer, which is consistent with increased chromosomal aberrations and hypermutability observed in cultured cells. These data suggest that each of these helicases contributes to maintaining genomic stability and that an important function of these helicases appears to be the resolution of recombination intermediates.

Introduction

In a wide range of organisms from the prokaryotes to eukaryotes, DNA helicases, with their ability to unwind DNA structures, are involved in many basic cellular processes such as replication, recombination, and repair. The RecQ family of helicases is a subfamily of the DExH-box-containing DNA helicases, which include *Escherichia coli* RecQ,[1] *Saccharomyces cerevisiae* Sgs1,[2,3] and *Schizosaccharomyces pombe* Rqh1.[4] In *E. coli* lacking a functional RecBCD helicase, *RecQ* mutations increase UV sensitivity.[1] RecQ was originally identified as a component of the RecF pathway of conjugational recombination operating in the absence of a functional RecBCD helicase.[1,5] Subsequent work showed that *E. coli* RecQ protein disrupts recombination intermediates to suppress the formation of large chromosomal deletions mediated by illegitimate recombination using short homologous sequences.[6]

Similar to bacteria, budding and fission yeasts have a single RecQ-like helicase. *SGS1* was initially found because its deficiency resulted in suppression of slow growth in *Top3* mutants of the budding yeast *S. cerevisiae*. In fact, SGS1 protein was found to interact with Top3 and to act as a 3'-to5' helicase.[2,7] In the fission yeast *S. pombe*, a *Rqh1* mutation was identified in radiaition-sensitive (*rad*) and hydroxyurea-sensitive (*hus*) mutants.[4] Genetic analysis demonstrated that *Rqh1* mutants are defective in the recombination bypass of UV-induced DNA damage during the S phase.[8] Like SGS1 protein, Rqh1 protein has a 3'-to5' DNA-unwinding activity and exists with Top3 in a high-molecular-weight complex in fission yeasts.[9]

DNA Repair and Human Disease, edited by Adayabalam S. Balajee. ©2006 Landes Bioscience and Springer Science+Business Media.

In humans, five homologues of the bacterial RecQ helicase have been identified. Disruption of their functions may reduce genomic stability and thus contribute to tumorigenesis.[10,11] Mutations in three different human RecQ helicase family members, the *WRN* gene, the *BLM* gene and the *RECQL4* gene, give rise to three distinct human disorders, Werner syndrome (WS), Bloom syndrome (BS), and Rothmund-Thomson syndrome (RTS), which lead to cancer predisposition and/or segmental premature aging.[12-14] Defining in vivo functions of RecQ family helicases is necessary for understanding the mechanisms of cancer predisposition and segmental premature aging in the *RecQ*-deficient disorders. The hyper-recombination phenotype of *RecQ*-deficient cells suggests that these helicases seem to maintain replication forks to ensure processive DNA replication, and/or carry out conventional recombination or recombination-mediated DNA repair. Much information obtained from previous studies on RecQ family helicases may illuminate the mechanism of action of these important molecules.

Bloom Syndrome Gene (*BLM*)

BS is characterized by severe pre- and post-natal growth retardation, immunodeficiency, sun-sensitive facial erythema, male infertility, genomic instability and predisposition to cancer in many types of tissue.[15] The varieties of cancer are almost identical to that seen in the general population, and also include Wilms tumor and osteosarcomas, which are considered more rare. Cultured cells from BS patients exhibit hypermutalism and chromosomal and DNA instability characterized by an excessive number of locus-specific mutations, including insertions, deletions, and loss of heterozygousity.[16] In additon, BS cells show a high frequency of microscopically visible chromaid gaps, breaks, and rearrangements, quadriradials, and associations at telomeres.[17]

The *BLM* gene is located on chromosome 15 at 15q26.1 and encodes a protein of 1417 a.a.[12] The *BLM* gene is not essential in humans, however, the murine homolog is essential in mice since the homozygous disruption of *BLM* results in embryonic lethality by E13.5.[18] Mice heterozygous for a targeted null mutation of *BLM* demonstrated that *BLM* haploinsufficiency enhances T cell tumorigenesis in response to murine leukemia virus infection and intestinal tumorigenesis when crossed with mice carrying a mutation in the *APC* tumor suppressor gene.[19]

Biochemical studies have shown that BLM has an ATP- and Mg^{2+}- dependent 3'-to-5' DNA-unwinding activity.[20] Like the SGS1 protein, the human BLM protein is a hexameric helicase.[21,22] Among the human RecQ homologues, BLM is most similar to two yeast RecQ-type helicases, SGS1 and Rqh1, both of which regulate genetic exchange and maintain genomic stability by preventing inappropriate recombination during interrupted S phase.[4,23,24] These three proteins contain a helicase domain of similar size and position as well as two acidic amino acid clusters in their amino termini.[3,4,12] In *S. cerevisiae*, *SGS1* mutants are hypersensitive to hydroxyurea (HU),[25] and they exhibit mitotic hyper-recombination, resulting in increased frequencies of ectopic, interchromosomal homologous exchange and intrachromosomal excision as well as poor sporulation.[23,24] *BLM* can suppress increased homologous and illegitimate recombination, and restore increased sensitivity to HU.[25] Also, *BLM*, but not *WRN*, can prevent premature aging and increased homologous recombination at the rDNA loci caused by *SGS1* mutation.[26]

Like the yeast *SGS1* mutants, an abnormally high incidence of spontaneous sister chromatid exchanges (SCEs) are observed in BS cells.[15,27,28] The increase in frequency of SCEs in BS cells is further facilitated by exposure to UVC light and DNA damaging agents such as ethyl methansulfonate, N-ethyl-N-nitrosourea and 5-bromodeoxyuridine (BrdU).[29-31] Although thymine dimers on single stranded DNA formed by UVC irradiation and base damage to genomic DNA caused by hydrolysis, oxidation and nonenzymatic methylation are usually repaired by the nucleotide-excision repair (NER) pathway prior to replication,[32] large populations of such DNA adducts may be overlooked by the NER pathway. When a replication fork is stalled at a DNA adduct, it may produce a single-strand gap between the damaged base and a new Okazaki fragment being synthesized downstream, possibly resulting in a chromatid break.[33,34] Such strand discontinuities can be repaired post-replicationally by homologous recombination (HR) with the sister chromatid, i.e. by SCE.[35] In fact, BS cells exhibit hypersensitivities to DNA damaging agents such as N-ethyl-N-nitrosourea, ethyl methanesulfonate, methyl methanesulfonate (MMS) and 4-nitroquinoline 1-oxide (4-NQO), as well as to irradiation with UVC.[36-38] In addition to BS cells, chicken *BLM* mutant DT40 cells exhibit hypersensitivity to UVC irradiation and genotoxic agents such as bleomycin, etoposide, MMS and 4-NQO.[39] They also show chromosomal aberrations that ultimately lead to cell death. Karyotypic analysis demonstrated that UVC irradiation during G_1 to early S phase causes chromosomal aberration prior to cell death in *BLM* mutant DT40 cells, suggesting that BLM is involved in early S phase-specific surveillance of damaged sites on DNA.

On the other hand, BLM has also been shown to possess the ability to selectively recognize Holliday junctions and efficiently promotes ATP-dependent branch migration of Holliday junctions during S phase in vitro, resulting in the prevention of promiscuous recombination events.[40] The level of SCE in *BLM* mutant DT40 cells is considerably reduced in the absence of *RAD54*, indicating that a large proportion of the SCE occurs via HR.[41] In addition, *BLM* mutant DT40 cells exhibit hyper-recombination as indicated by the increased frequency of targeted genome integrations in addition to that of SCE.[39,41] Thus, a *BLM* defect would lead to the initiation of double-strand break (DSB) repair by HR, resulting in an increase in frequency of SCE. DNA crosslinking within GC-rich regions exposed by recombination or replication in mammalian chromosomes can create quadriradials which are well described as one of the representative phenotypes seen in BS cells.[42] Previous work has shown that BLM is a general helicase with preference for a quadriradial formed in a single stranded G-rich region exposed by recombination.[43] The fact that the number of chromatid quadriradials observed in the *BLM* mutant DT40 cells are elevated after UVC irradiation supports this speculation.[39]

BLM localizes to promyelocytic leukemia protein (PML) nuclear bodies,[44] and its expression is cell cycle-regulated, peaking in late S phase and G_2.[45,46] Both the recombination/repair proteins hRAD51 and replication protein (RP)-A assemble with BLM in PML nuclear bodies during late S/G_2 phase. BS cells have been reported to be more sensitive than wild-type cells to radiation, particularly in late S and G_2 phase.[47,48] BLM protein assembles into foci specifically in response to agents that cause DNA DSBs in normal cells, but not in cells with defective PML.[44] HR repair is carried out by the RAD52 complex, which includes hRAD51 and RP-A.[49,50] These studies suggest that BLM is part of a nuclear matrix-based complex that requires PML and functions in HR repair after DNA damage. Additionally, BLM also has been identified as a member of a group of proteins that associate with BRCA1 to form a large complex that includes tumor suppressors and DNA damage repair proteins MSH2, MSH6, MLH1, ATM, and the RAD50-MRE11-NBS1 protein complex.[51] Recently, It was observed that BLM is required for correct nuclear localization of RAD50-MRE11-NBS1 complexes after replication fork arrest.[52] The γ-irradiation-induced BLM phosphorylation requires functional ATM.[53] Both UVC radiation and HU-mediated DNA synthesis inhibition induced BLM phosphorylation via an ATM-independent pathway.[54] Among the assembly components of the BRCA1-associated large complex, both BRCA1 and NBS1 are phosphorylated via an ATM-dependent pathway in response to ionizing radiation, and via an ATM-independent pathway in response to UVC and HU.[55,56] It has been demonstrated that ATM responds exclusively to DNA DSBs, wheras the ATM-related kinase, ATR, exhibits substrate specificity similar to ATM and also reacts to UV damage and replication arrest.[57,58] The BRCA1 phosphorylation induced by UVC or HU has been shown to be controlled by ATR.[56] These findings suggest that ATR phosphorylates BLM upon UV damage and replication arrest. Damage during S phase causes BLM to rapidly form nuclear foci at replication forks that develop DNA breaks. These BLM foci then recruit BRCA1 and NBS1. Assembly of BRCA1 and NBS1 repair complxes at stalled replication forks was markedly delayed in damaged BS cells.[59] It has been suggested that BLM plays a role in repair by recognizing DNA adducts and recruiting BRCA1 to the damage-induced nuclear matrix, which results in assembly of the BRCA-associated repair complex at a stalled replication fork. Concurrently or alternatively, BLM functions in HR repair after DNA damage as part of a RAD52 complex, and/or by resolving DNA structures such as Holliday junctions at stalled replication forks in addition to quadriradial structures prior to HR repair.

Werner Syndrome Gene (WRN)

Individuals afflicted with WS exhibit at an early age, features associated with the normal advanced aging process such as bilateral cataracts, diabetes mellitus, atrophy of the skin, graying and loss of hair, osteoporosis, atherosclerosis and increased predisposition to development of cancer.[60] The most prevalent types of cancer are soft tissue sarcomas, however, thyroid carcinomas, osteosarcomas, meningiomas, and melanomas are also seen. In vitro studies of fibroblast growth characteristics suggest that WS may be related to normal aging. The life span of WS fibroblasts as measured by in vitro population doubling levels is much shorter than that of normal fibroblasts, and they have a prolonged S phase.[61,62] WS cells are also characterized by genomic instability in the form of variegated translocation mosaicism and extensive deletions.[63,64] WS cells display a defect in the G_2-phase decatenation checkpoint which is known to inhibit progression into mitosis until chromatids are correctly decatenated by topoisomerase II.[65] In WS cells, failure to phosphorylate BRCA1 in an ATR-dependent manner in response to decatenation checkpoint activation results in enhanced chromosomal damage and apoptosis.[65] It has been shown that the suppression of RAD51-dependent recombination leads to significantly improved survival of WS cells following DNA damage, suggesting defective recombination resolution in WS cells.[66] Like *BLM*, *WRN* can suppress increased homologous and illegitimate recombination in the *SGS1* mutant.[25]

The *WRN* gene is located on chromosome 8 at 8p11-12 and encodes a protein of 1432 a.a..[13,67] Subcellular localization studies revealed that WRN protein is localized both in the nucleoplasm and the nucleoli.[68] Transcriptional activation facilitates translocation of WRN protein from the nucleoplasm to the nucleoli,[69] whereas DNA damages lead to extensive translocation of WRN from the nucleolus to nucleoplasmic foci.[70] In addition, WRN is acetylated in vivo, which is markedly stimulated by the acetyltransferase p300, which results in the augmentation of WRN translocation into nucleoplasmic foci.[70] WRN has DNA-dependent ATPase activities and a 3'-to-5' unwinding activity for not only duplex DNA but also RNA-DNA heteroduplexes.[71] A region near the N-terminus of WRN contains three conserved motifs that resemble the conserved motifs in the proofreading exonuclease domain of *E. coli* DNA polymerase I and in *E. coli* RNaseD.[72] In fact, WRN protein has an exonuclease activity although there is some disagreement on its directionality and dependency on helicase activity.[73-76] WRN exonuclease preferentially hydrolyzes alternative DNA structures that contain bubbles, extra-helical loops, or 3-way or 4-way junctions.[77,78] Like BLM, WRN can resolve aberrant DNA structures such as G-quadruplex and G-triplex DNAs.[43,79]

WRN-null mouse mutants are born at the expected Mendelian frequency and are healthy and fertile, showing no signs of premature organismic aging or increased rates of tumor formation. Cells from these animals do not show elevated susceptibility to genotoxins. Thus, the knockout of the *WRN* gene in mice does not recapitulate many of the phenotypes of human WS. It is possible that defective nucleolar localization of murine WRN protein may be related to this discrepancy of phenotypes.[68] However, mice lacking *WRN* display a shorter life span in a *p53*[-/-] background.[80]

WS cells exhibit a delayed and attenuated accumulation of p53 after exposure to UV irradiation.[81] Notably, WRN contributes to the induction of p53 by various DNA damaging agents. WRN and p53 can form specific protein-protein interactions through their respective C-terminal domains.[82] In addition, WRN has been shown to bind to and/or functionally interact with RPA,[83,84] proliferating cell nuclear antigen (PCNA), and DNA topoisomerase I. [85] Each of these interacting proteins is involved in DNA manipulations including those that resolve alternative DNA structures. WRN also interacts functionally with polymerase δ which is a eukaryotic polymerase required for DNA replication and repair.[86] WRN is phosphorylated in vitro by DNA-PK and requires DNA-PK for its phosphorylation in vivo.[87] WRN interacts directly with the catalytic subunit of DNA-PK (DNA-PK$_{CS}$), which inhibits both helicase and exonuclease activities of WRN. In addition, WRN forms a stable complex on DNA with DNA-PK$_{CS}$ and Ku 86/70.[87] Although Ku proteins have no effect on ATPase or helicase activity, they strongly stimulate the specific exonuclease activity of WRN.[88] When WRN is assembled with DNA-PK$_{CS}$ and Ku proteins, WRN enzymatic inhibition by DNA-PK$_{CS}$ is reversed. In response to replication blockage, WRN is phosphorylated in an ATR/ATM-dependent manner and colocalizes with ATR, suggesting that WRN and ATR kinase collaborate to prevent genomic instability during S phase.[89] WRN also forms a complex with RAD52 in vivo that colocalizes in foci associated with arrested replication forks.[90] WRN has been known to promote the ATP-dependent translocation of Holliday junctions, an activity that is also exhibited by BLM.[84] WRN increases the efficiency of RAD52-mediated strand annealing between non-duplex DNA and homologous sequences, whereas RAD52 can both inhibit and enhance WRN helicase activity in a DNA structure-dependent manner.[90]

Hypersensitivities of B-lymphoblastoid cells from WS patients to 4-nitroquinoline 1-oxide (4-NQO), etoposide and camptothecin (CPT) have been reported.[91-93] However, WS B-lymphoblastoid cells do not exhibit a hypersensitivity to other DNA-damaging agents such as most alkylating agents, and X-rays, bleomycin, or H_2O_2 that produce reactive oxygen species,[91,94,95] as well as to UV irradiation.[96] In chicken *WRN* mutant DT40 cells irradiated with UV light, the frequency of chromatid breaks are identical to that observed in wild-type cells.[97] In *WRN/BLM*-double knockout DT40 cells, UV irradiation in the late S to G_2 phase synergistically enhances the increase in the number of chromatid breaks.[97] *WRN*-mutant DT40 cells also exhibit an increased incidence of spontaneous HR. However, disruption of *WRN* partially diminishes the SCE frequency which is increased in *BLM*-mutant DT40 cells despite the fact that the SCE frequency does not change in *WRN*-mutant cells compared with wild-type cells.[97] These results imply that WRN may contribute to accelerate the post-replicational HR repair which occurs due to the absence of BLM.

Rothmund-Thomson Syndrome Gene (RT)

Rothmund-Thomson syndrome (RTS) patients exhibit a heterogeneous clinical profile that includes a characteristic skin rash (poikiloderma), small stature, sparse hair, bony abnormalities, juvenile cataracts, and an increased risk of cancer, particularly osteosarcoma, a malignant primary bone tumor.[98] Several cases of RTS patients with frequent infections, impaired lymphocyte function, and decreased T lymphocyte and leukocyte counts have been reported, although distinct immunological dysfunction has not been observed in these patients.[99] A pure candidate gene approach showed that mutations in the *RECQL4* gene, which is located on human chromosome 8q24.3, occurred in at least some cases of RTS.[14] Neither complementation nor linkage studies have been reported, suggesting that mutations in more than one gene are responsible for RTS. In addition, the absence of detectable mutations in the *RECQL4* gene in approximately one-third of the patients has been reported in previous studies.[14,100] Thus, the finding of genetic heterogeneity in RTS contrasts with BS and WS.

RECQL4-knockout mice in which exons 5-8 were disrupted, die between embryonic day 3.5-6.5.[101] The growth rate of both the inner cell mass and trophoblast cells of the blastocysts from the *RECQL4*-deficient mice markedly decrease. Recent studies demonstrated that the exon 13-deleted *RECQL4*-deficient mice are viable, but exhibit severe growth retardation and abnormalities in several tissues, and that embryonic fibroblasts show a defect in cell proliferation.[102] These abnormalities in the *RECQL4*-deficient mice are similar to those exhibited in RTS patients. Exon 13 is one of the coding exons of the consensus RecQ-helicase domain. This domain is the primary site of mutations that have been identified in RTS patients.

In addition to WS cells and BS cells, RTS cells show genomic instability, including trisomy, aneuploidy and chromosomal rearrangements. However, in some cases of RTS, chromosomal breakage in lymphocytes is not notably greater than that of unaffected controls.[103-105] Conflicting studies report that cells from RTS patients exhibit both increased[106,107] or no change[108,109] in sensitivity to UV and γ irradiation. Embryonic fibroblasts derived from exon 13-deleted *RECQL4*-deficient mice do not exhibit a statistically significant difference in sensitivity to UV and ionizing radiation in comparison with wild type cells.[102] RTS cells exhibit normal sister chromatid exchange rates and a normal response to genotoxin-induced DNA breakage in vitro.[105] There has been comparatively little information available on the RECQL4 protein and proteins interacting with it, thus, functions of RECQL4 protein concerning DNA repair remain unclear.

Acknowledgements

I would like to thank Leslie James R. Bestilny at RIKEN Yokohama Institute for his critical reading of the manuscript.

References

1. Nakayama K, Irino N, Nakayama H. The RecQ gene of Escherichia coli K12: Molecular cloning and isolation of insertion mutants. Mol Gen Genet 1985; 200:266-271.
2. Gangloff S, McDonald JP, Bendixen C et al. The yeast type I topoisomerase Top3 interacts with Sgs1, a DNA helicase homolog: A potential eukaryotic reverse gyrase. Mol Cell Biol 1994; 14:8391-8398.
3. Watt PM, Louis EJ, Borts RH et al. Sgs1: A eukaryotic homolog of E. coli RecQ that interacts with topoisomerase II in vivo and is required for faithful chromosome segregation. Cell 1995; 81(2):253-260.
4. Stewart E, Chapman CR, Al-Khodairy F et al. rqh1+, a fission yeast gene related to the Bloom's and Werner's syndrome genes, is required for reversible S phase arrest. EMBO J 1997; 16:2682-2692.
5. Nakayama H, Nakayama K, Nakayama R et al. Isolation and genetic characterization of a thymineless death-resistant mutant of Escherichia coli K12: Identification of a new mutation (recQ1) that blocks the RecF recombination pathway. Mol Gen Genet 1984; 195:474-480.
6. Hanada K, Ukita T, Kohno Y et al. RecQ DNA helicase is a suppressor of illegitimate recombination in Escherichia coli. Proc Natl Acad Sci USA 1997; 94:3860-3865.
7. Bennett RJ, Noirot-Gros MF, Wang JC. Interaction between yeast sgs1 helicase and DNA topoisomerase III. J Biol Chem 2000; 275:26898-26905.
8. Murray JM, Lindsay HD, Munday CA et al. Role of schizosaccharomyces pombe RecQ homolog, recombination, and checkpoint genes in UV damage tolerance. Mol Cell Biol 1997; 17:6868-6875.
9. Laursen LV, Bjergbaek L, Murray JM et al. RecQ helicases and topoisomerase III in cancer and aging. Biogerontology 2003; 4:275-287.
10. van Brabant AJ, Stan R, Ellis NA. DNA helicases, genomic instability, and human genetic disease. Annu Rev Genomics Hum Genet 2000; 1:409-459.
11. Karow JK, Wu L, Hickson ID. RecQ family helicases: Roles in cancer and aging. Curr Opin Genet Dev 2000; 10:32-38.
12. Ellis NA, Groden J, Ye TZ et al. The Bloom's syndrome gene product is homologous to RecQ helicases. Cell 1995; 83:655-666.

13. Yu CE, Oshima J, Fu YH et al. Positional cloning of the Werner's syndrome gene. Science 1996; 272:258-262.
14. Kitao S, Shimamoto A, Goto M et al. Mutations in RECQL4 cause a subset of cases of Rothmund-Thomson syndrome. Nat Genet 1999; 22:82-84.
15. German J. Bloom syndrome: A mendelian prototype of somatic mutational disease. Medicine (Baltimore). 1993; 72:393-406.
16. Rosin MP, German J. Evidence for chromosome instability in vivo in Bloom syndrome: Increased numbers of micronuclei in exfoliated cells. Hum Genet 1985; 71:187-191.
17. Chakraverty RK, Hickson ID. Defending genome integrity during DNA replication: A proposed role for RecQ family helicases. Bioessays 1999; 21:286-294.
18. Chester N, Kuo F, Kozak C et al. Stage-specific apoptosis, developmental delay, and embryonic lethality in mice homozygous for a targeted disruption in the murine Bloom's syndrome gene. Genes Dev 1998; 12:3382-3393.
19. Goss KH, Risinger MA, Kordich JJ et al. Enhanced tumor formation in mice heterozygous for Blm mutation. Science 2002; 297:2051-2053.
20. Karow JK, Chakraverty RK, Hickson ID. The Bloom's syndrome gene product is a 3'-5' DNA helicase. J Biol Chem 1997; 272:30611-30614.
21. Bennett RJ, Sharp JA, Wang JC. Purification and characterization of the Sgs1 DNA helicase activity of Saccharomyces cerevisiae. J Biol Chem 1998; 273:9644-9650.
22. Karow JK, Newman RH, Freemont PS et al. Oligomeric ring structure of the Bloom's syndrome helicase. Curr Biol 1999; 9:597-600.
23. Davey S, Han CS, Ramer SA et al. Fission yeast rad12+ regulates cell cycle checkpoint control and is homologous to the Bloom's syndrome disease gene. Mol Cell Biol 1998; 18:2721-2728.
24. Watt PM, Hickson ID, Borts RH et al. SGS1, a homologue of the Bloom's and Werner's syndrome genes, is required for maintenance of genome stability in Saccharomyces cerevisiae. Genetics 1996; 144:935-945.
25. Yamagata K, Kato J, Shimamoto A et al. Bloom's and Werner's syndrome genes suppress hyperrecombination in yeast sgs1 mutant: Implication for genomic instability in human diseases. Proc Natl Acad Sci USA 1998; 95:8733-8738.
26. Heo SJ, Tatebayashi K, Ohsugi I et al. Bloom's syndrome gene suppresses premature ageing caused by Sgs1 deficiency in yeast. Genes Cells 1999; 4:619-625.
27. Ray JH, German J. Bloom's syndrome and EM9 cells in BrdU-containing medium exhibit similarly elevated frequencies of sister chromatid exchange but dissimilar amounts of cellular proliferation and chromosome disruption. Chromosoma 1984; 90:383-388.
28. Chaganti RS, Schonberg S, German J. A manyfold increase in sister chromatid exchanges in Bloom's syndrome lymphocytes. Proc Natl Acad Sci USA 1974; 71:4508-4512.
29. Kurihara T, Tatsumi K, Takahashi H et al. Sister-chromatid exchanges induced by ultraviolet light in Bloom's syndrome fibroblasts. Mutat Res 1987; 183:197-202.
30. Heartlein MW, Tsuji H, Latt SA. 5-Bromodeoxyuridine-dependent increase in sister chromatid exchange formation in Bloom's syndrome is associated with reduction in topoisomerase II activity. Exp Cell Res 1987; 169:245-254.
31. Krepinsky AB, Rainbow AJ, Heddle JA. Studies on the ultraviolet light sensitivity of Bloom's syndrome fibroblasts. Mutat Res 1980; 69:357-368.
32. Wood RD. Repair of pyrimidine dimer ultraviolet light photoproducts by human cell extracts. Biochemistry 1989; 28:8287-8292.
33. Sonoda E, Sasaki MS, Morrison C et al. Sister chromatid exchanges are mediated by homologous recombination in vertebrate cells. Mol Cell Biol 1999; 19:5166-5169.
34. Yamaguchi-Iwai Y, Sonoda E, Sasaki MS et al. Mre11 is essential for the maintenance of chromosomal DNA in vertebrate cells. EMBO J 1999; 18:6619-6629.
35. Kadyk LC, Hartwell LH. Replication-dependent sister chromatid recombination in rad1 mutants of Saccharomyces cerevisiae. Genetics 1993; 133:469-487.
36. Krepinsky AB, Heddle JA, German J. Sensitivity of Bloom's syndrome lymphocytes to ethyl methanesulfonate. Hum Genet 1979; 50:151-156.
37. Kurihara T, Inoue M, Tatsumi K. Hypersensitivity of Bloom's syndrome fibroblasts to N-ethyl-N-nitrosourea. Mutat Res 1987; 184:147-151.

38. Shiraishi Y. Bloom syndrome B-lymphoblastoid cells are hypersensitive towards carcinogen and tumor promoter-induced chromosomal alterations and growth in agar. EMBO J 1985; 4:2553-2560.

39. Imamura O, Fujita K, Shimamoto A et al. Bloom helicase is involved in DNA surveillance in early S phase in vertebrate cells. Oncogene 2001; 20:1143-1151.

40. Karow JK, Constantinou A, Li JL et al. The Bloom's syndrome gene product promotes branch migration of holliday junctions. Proc Natl Acad Sci USA 2000; 97:6504-6508.

41. Wang W, Seki M, Narita Y et al. Possible association of BLM in decreasing DNA double strand breaks during DNA replication. EMBO J 2000; 19:3428-3435.

42. Werner-Favre C, Wyss M, Cabrol C et al. Cytogenetic study in a mentally retarded child with Bloom syndrome and acute lymphoblastic leukemia. Am J Med Genet 1984; 18:215-221.

43. Sun H, Karow JK, Hickson ID et al. The Bloom's syndrome helicase unwinds G4 DNA. J Biol Chem 1998; 273:27587-27592.

44. Bischof O, Kim SH, Irving J et al. Regulation and localization of the Bloom syndrome protein in response to DNA damage. J Cell Biol 2001; 153:367-380.

45. Gharibyan V, Youssoufian H. Localization of the Bloom syndrome helicase to punctate nuclear structures and the nuclear matrix and regulation during the cell cycle: Comparison with the Werner's syndrome helicase. Mol Carcinog 1999; 26:261-273.

46. Dutertre S, Ababou M, Onclercq R et al. Cell cycle regulation of the endogenous wild type Bloom's syndrome DNA helicase. Oncogene 2000; 19:2731-2738.

47. Aurias A, Antoine JL, Assathiany R et al. Radiation sensitivity of Bloom's syndrome lymphocytes during S and G2 phases. Cancer Genet Cytogenet 1985; 16:131-136.

48. Hall EJ, Marchese MJ, Astor MB et al. Response of cells of human origin, normal and malignant, to acute and low dose rate irradiation. Int J Radiat Oncol Biol Phys 1986; 12:655-659.

49. Kanaar R, Hoeijmakers JH, van Gent DC. Molecular mechanisms of DNA double strand break repair. Trends Cell Biol 1998; 8:483-489.

50. Thompson LH, Schild D. The contribution of homologous recombination in preserving genome integrity in mammalian cells. Biochimie 1999; 81:87-105.

51. Wang Y, Cortez D, Yazdi P et al. BASC, a super complex of BRCA1-associated proteins involved in the recognition and repair of aberrant DNA structures. Genes Dev 2000; 14:927-939.

52. Franchitto A, Pichierri P. Bloom's syndrome protein is required for correct relocalization of RAD50/MRE11/NBS1 complex after replication fork arrest. J Cell Biol 2002; 157:19-30.

53. Ababou M, Dutertre S, Lecluse Y et al. ATM-dependent phosphorylation and accumulation of endogenous BLM protein in response to ionizing radiation. Oncogene 2000; 19:5955-5963.

54. Ababou M, Dumaire V, Lecluse Y et al. Cleavage of BLM and sensitivity of Bloom's syndrome cells to hydroxurea and UV-C radiation. Cell Cycle 2002; 1:262-266.

55. Wu L, Davies SL, North PS et al. The Bloom's syndrome gene product interacts with topoisomerase III. J Biol Chem 2000; 275:9636-9644.

56. Tibbetts RS, Cortez D, Brumbaugh KM et al. Functional interactions between BRCA1 and the checkpoint kinase ATR during genotoxic stress. Genes Dev 2000; 14:2989-3002.

57. Kim ST, Lim DS, Canman CE et al. Substrate specificities and identification of putative substrates of ATM kinase family members. J Biol Chem 1999; 274:37538-37543.

58. Shiloh Y. ATM (ataxia telangiectasia mutated): Expanding roles in the DNA damage response and cellular homeostasis. Biochem Soc Trans 2001; 29:661-666.

59. Davalos AR, Campisi J. Bloom syndrome cells undergo p53-dependent apoptosis and delayed assembly of BRCA1 and NBS1 repair complexes at stalled replication forks. J Cell Biol 2003; 162:1197-1209.

60. Epstein CJ. Werner's syndrome and aging: A reappraisal. Adv Exp Med Biol 1985; 190:219-228.

61. Faragher RG, Kill IR, Hunter JA et al. The gene responsible for Werner syndrome may be a cell division "counting" gene. Proc Natl Acad Sci USA 1993; 90:12030-12034.

62. Salk D, Bryant E, Hoehn H et al. Growth characteristics of Werner syndrome cells in vitro. Adv Exp Med Biol 1985; 190:305-311.

63. Salk D, Au K, Hoehn H et al. Cytogenetic aspects of Werner syndrome. Adv Exp Med Biol 1985; 190:541-546.

64. Fukuchi K, Martin GM, Monnat Jr RJ et al. Mutator phenotype of Werner syndrome is characterized by extensive deletions. Proc Natl Acad Sci USA 1989; 86:5893-5897.

65. Franchitto A, Oshima J, Pichierri P. The G2-phase decatenation checkpoint is defective in Werner syndrome cells. Cancer Res 2003; 63:3289-3295.

66. Saintigny Y, Makienko K, Swanson C et al. Homologous recombination resolution defect in werner syndrome. Mol Cell Biol 2002; 22:6971-6978.

67. Goto M, Rubenstein M, Weber J et al. Genetic linkage of Werner's syndrome to five markers on chromosome 8. Nature 1992; 355:735-738.

68. Suzuki T, Shiratori M, Furuichi Y et al. Diverged nuclear localization of Werner helicase in human and mouse cells. Oncogene 2001; 20:2551-2558.

69. Shiratori M, Suzuki T, Itoh C et al. WRN helicase accelerates the transcription of ribosomal RNA as a component of an RNA polymerase I-associated complex. Oncogene 2002; 21:2447-2454.

70. Blander G, Zalle N, Daniely Y et al. DNA damage-induced translocation of the Werner helicase is regulated by acetylation. J Biol Chem 2002; 277:50934-50940.

71. Suzuki N, Shimamoto A, Imamura O et al. DNA helicase activity in Werner's syndrome gene product synthesized in a baculovirus system. Nucleic Acids Res 1997; 25:2973-2978.

72. Mushegian AR, Bassett Jr DE, Boguski MS et al. Positionally cloned human disease genes: Patterns of evolutionary conservation and functional motifs. Proc Natl Acad Sci USA 1997; 94:5831-5836.

73. Huang S, Li B, Gray MD et al. The premature ageing syndrome protein, WRN, is a 3'—>5' exonuclease. Nat Genet 1998; 20:114-116.

74. Kamath-Loeb AS, Shen JC, Loeb LA et al. Werner syndrome protein. II. Characterization of the integral 3' —> 5' DNA exonuclease. J Biol Chem 1998; 273:34145-34150.

75. Shen JC, Gray MD, Oshima J et al. Werner syndrome protein. I. DNA helicase and dna exonuclease reside on the same polypeptide. J Biol Chem 1998; 273:34139-34144.

76. Suzuki N, Shiratori M, Goto M et al. Werner syndrome helicase contains a 5'—>3' exonuclease activity that digests DNA and RNA strands in DNA/DNA and RNA/DNA duplexes dependent on unwinding. Nucleic Acids Res 1999; 27:2361-2368.

77. Machwe A, Ganunis R, Bohr VA et al. Selective blockage of the 3'—>5' exonuclease activity of WRN protein by certain oxidative modifications and bulky lesions in DNA. Nucleic Acids Res 2000; 28:2762-2670.

78. Shen JC, Loeb LA. Werner syndrome exonuclease catalyzes structure-dependent degradation of DNA. Nucleic Acids Res 2000; 28:3260-3268.

79. Brosh Jr RM, Majumdar A, Desai S et al. Unwinding of a DNA triple helix by the Werner and Bloom syndrome helicases. J Biol Chem 2001; 276:3024-3030.

80. Lombard DB, Beard C, Johnson B et al. Mutations in the WRN gene in mice accelerate mortality in a p53-null background. Mol Cell Biol 2000; 20:3286-3291.

81. Blander G, Zalle N, Leal JF et al. The Werner syndrome protein contributes to induction of p53 by DNA damage. FASEB J 2000; 14:2138-2140.

82. Blander G, Kipnis J, Leal JF et al. Physical and functional interaction between p53 and the Werner's syndrome protein. J Biol Chem 1999; 274:29463-29469.

83. Brosh Jr RM, Orren DK, Nehlin JO et al. Functional and physical interaction between WRN helicase and human replication protein A. J Biol Chem 1999; 274:18341-18350.

84. Constantinou A, Tarsounas M, Karow JK et al. Werner's syndrome protein (WRN) migrates Holliday junctions and co-localizes with RPA upon replication arrest. EMBO Rep 2000; 1:80-84.

85. Lebel M, Spillare EA, Harris CC et al. The Werner syndrome gene product co-purifies with the DNA replication complex and interacts with PCNA and topoisomerase I. J Biol Chem 1999; 274:37795-37799.

86. Kamath-Loeb AS, Johansson E, Burgers PM et al. Functional interaction between the Werner Syndrome protein and DNA polymerase delta. Proc Natl Acad Sci USA 2000; 97:4603-4608.

87. Yannone SM, Roy S, Chan DW et al. Werner syndrome protein is regulated and phosphorylated by DNA-dependent protein kinase. J Biol Chem. 2001; 276:38242-38248.

88. Cooper MP, Machwe A, Orren DK et al. Ku complex interacts with and stimulates the Werner protein. Genes Dev 2000; 14:907-912.

89. Pichierri P, Rosselli F, Franchitto A. Werner's syndrome protein is phosphorylated in an ATR/ATM-dependent manner following replication arrest and DNA damage induced during the S phase of the cell cycle. Oncogene 2003; 22:1491-1500.

90. Baynton K, Otterlei M, Bjoras M et al. WRN interacts physically and functionally with the recombination mediator protein RAD 52. J Biol Chem 2003; 278:36476-36486.

91. Gebhart E, Bauer R, Raub U et al. Spontaneous and induced chromosomal instability in Werner syndrome. Hum Genet 1988; 80:135-139.

92. Ogburn CE, Oshima J, Poot M et al. An apoptosis-inducing genotoxin differentiates heterozygotic carriers for Werner helicase mutations from wild-type and homozygous mutants. Hum Genet 1997; 101:121-125.

93. Elli R, Chessa L, Antonelli A et al. Effects of topoisomerase II inhibition in lymphoblasts from patients with progeroid and "chromosome instability" syndromes. Cancer Genet Cytogenet 1996; 87:112-116.

94. Fujiwara Y, Higashikawa T, Tatsumi M. A retarded rate of DNA replication and normal level of DNA repair in Werner's syndrome fibroblasts in culture. J Cell Physiol. 1977; 92:365-374.

95. Okada M, Goto M, Furuichi Y et al. Differential effects of cytotoxic drugs on mortal and immortalized B-lymphoblastoid cell lines from normal and Werner's syndrome patients. Biol Pharm Bull 1998; 21:235-239.

96. Higashikawa T, Fujiwara Y. Normal level of unscheduled DNA synthesis in Werner's syndrome fibroblasts in culture. Exp Cell Res 1978; 113:438-442.

97. Imamura O, Fujita K, Itoh C et al. Werner and Bloom helicases are involved in DNA repair in a complementary fashion. Oncogene 2002; 21:954-963.

98. Wang LL, Levy ML, Lewis RA et al. Clinical manifestations in a cohort of 41 Rothmund-Thomson syndrome patients. Am J Med Genet 2001; 102:11-17.

99. Vennos EM, James WD. Rothmund-Thomson syndrome. Dermatol Clin 1995; 13:143-150.

100. Wang LL, Gannavarapu A, Kozinetz CA et al. Association between osteosarcoma and deleterious mutations in the RECQL4 gene in Rothmund-Thomson syndrome. J Natl Cancer Inst 2003; 95:669-674.

101. Ichikawa K, Noda T, Furuichi Y. Preparation of the gene targeted knockout mice for human premature aging diseases, Werner syndrome, and Rothmund-Thomson syndrome caused by the mutation of DNA helicases. Nippon Yakurigaku Zasshi 2002; 119:219-226.

102. Hoki Y, Araki R, Fujimori A et al. Growth retardation and skin abnormalities of the Recql4-deficient mouse. Hum Mol Genet 2003; 12:2293-2299.

103. Der Kaloustian VM, McGill JJ, Vekemans M et al. Clonal lines of aneuploid cells in Rothmund-Thomson syndrome. Am J Med Genet 1990; 37:336-339.

104. Kerr B, Ashcroft GS, Scott D et al. Rothmund-Thomson syndrome: Two case reports show heterogeneous cutaneous abnormalities, an association with genetically programmed ageing changes, and increased chromosomal radiosensitivity. J Med Genet 1996; 33:928-934.

105. Lindor NM, Devries EM, Michels VV et al. Rothmund-Thomson syndrome in siblings: Evidence for acquired in vivo mosaicism. Clin Genet 1996; 49:124-129.

106. Shinya A, Nishigori C, Moriwaki S et al. A case of Rothmund-Thomson syndrome with reduced DNA repair capacity. Arch Dermatol 1993; 129:332-336.

107. Smith PJ, Paterson MC. Enhanced radiosensitivity and defective DNA repair in cultured fibroblasts derived from Rothmund Thomson syndrome patients. Mutat Res 1982; 94:213-228.

108. Starr DG, McClure JP, Connor JM. Non-dermatological complications and genetic aspects of the Rothmund-Thomson syndrome. Clin Genet 1985; 27:102-104.

109. Abrahams PJ, Houweling A, Schouten R et al. Abnormal kinetics of induction of UV-stimulated recombination in human DNA repair disorders. DNA Repair (Amst) 2003; 2:1211-1225.

CHAPTER 3

Trichothiodystrophy:

A Disorder Highlighting the Crosstalk between DNA Repair and Transcription

Miria Stefanini

Abstract

Trichothiodystrophy (TTD) is a rare autosomal recessive multisystem disorder characterized by sulfur-deficient brittle hair, mental and physical retardation, ichthyosis, and, in many patients, cutaneous photosensitivity but no cancer incidence. All sun-sensitive TTD cases appear to be defective in nucleotide excision repair (NER) as a consequence of alterations in one of three genes, namely *XPB*, *XPD* and *TTDA*. Intriguingly, in view of the very marked differences in the clinical phenotypes, defects in two of the genes altered in TTD (*XPB* and *XPD*) can also cause the cancer-prone disorder xeroderma pigmentosum (XP) or, in rare cases, the combined symptoms of XP and Cockayne syndrome (XP/CS). A breakthrough in understanding the perplexing features of this complex triad of hereditary disorders came from the discovery that the genes mutated in TTD are all related to TFIIH, a multiprotein complex involved in both transcription and NER. *XPB* and *XPD* encode two subunits of TFIIH, and also TTDA has been recently identified as a new component of the TFIIH complex. The discovery of this unexpected link between DNA repair and transcription was crucial to rationalize the TTD pathological phenotype as well as the puzzling genotype-phenotype relationships related to defects in *XPB* and *XPD*. The last few years have witnessed significant progress in this field. It was shown that the mutations associated with the three disorders are located at different sites in the *XPD* gene and that all the genetic and molecular alterations responsible for the NER-defective form of TTD cause a decrease by up to 70% in the cellular concentration of TFIIH. This implies that a limited availability of TFIIH interferes with NER but is compatible with life and therefore its impact on transcription must be selective only under certain conditions or in selective cellular compartments. We are also beginning to understand how different mutations in the *XPD* gene result in different clinical entities: all the *XPD* mutations found in patients, independent of the associated phenotype, are detrimental for the XPD helicase activity, thus explaining the NER defect, but only those responsible for the TTD phenotype affect basal transcription. Emerging evidence indicates that the involvement of TFIIH in transcription is multifaceted, ranging from transcription by RNA polymerase I and II to regulation of gene expression. Although the multiple roles of TFIIH in transcription remain to be fully explored, these discoveries represent major advances in our understanding of fundamental cellular processes. They have important impacts on clinical medicine with implications for cancer prevention, aging, differentiation and development.

DNA Repair and Human Disease, edited by Adayabalam S. Balajee. ©2006 Landes Bioscience and Springer Science+Business Media.

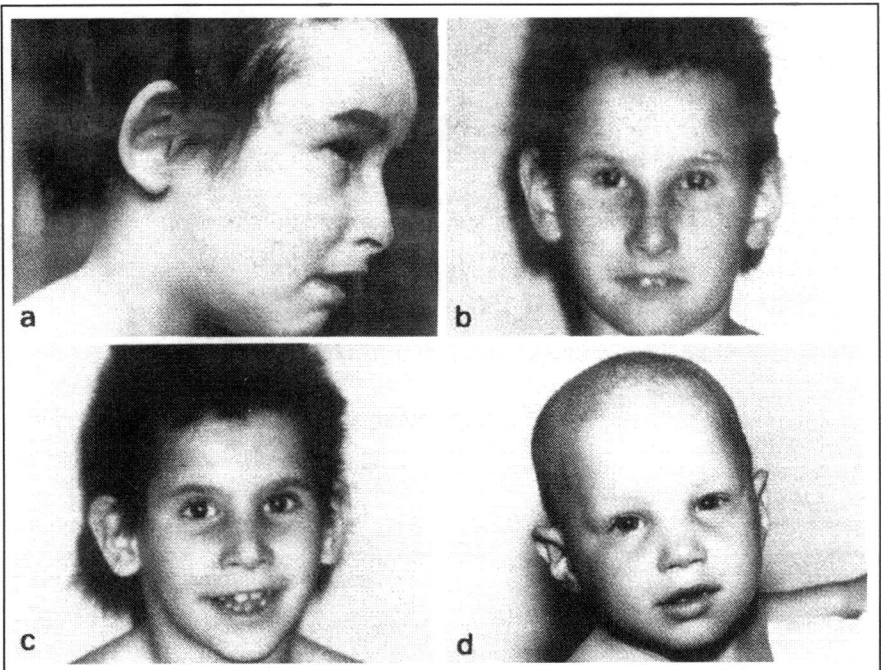

Figure 1. The four Italian patients affected by trichothiodystrophy, firstly described as NER-defective and classified into the XP-D group. From reference 3, with permission.

Introduction

Trichothiodystrophy (TTD), variously known as Tay syndrome, Pollit syndrome, Amish hair-brain syndrome and Sabinas syndrome, is a rare autosomal recessive disorder that was first described as a distinct clinical entity in 1980 to characterize the condition of patients with sulfur-deficient brittle hair and other neuroectodermal symptoms and signs.[1] The identification of DNA repair defects in a consistent number of TTD cases has stimulated further investigations aimed at clarifying the pathogenesis of this disorder whose clinical symptoms are more easily attributable to transcription anomalies than DNA repair alterations.

Clinical Features

The main diagnostic criteria of TTD are: brittle hair, mental and growth retardation, face characterized by receding chin, small nose, large ears and microcephaly, nail dysplasia and ichthyosis (Fig. 1). The acronyms PIBIDS, IBIDS, BIDS, have been used on the basis of the presence or absence of the following symptoms: Photosensitivity, Ichthyosis, Brittle hair, Impaired intelligence, Decreased fertility and Short stature. At birth, children often present with ichthyosiform erythroderma, and they may be encased in a collodion-like membrane. In addition, numerous patients suffer from repeated and severe infectious illnesses, mainly of the gastrointestinal and respiratory tract. Forty to 50% of patients exhibit marked photosensitivity but there are no reports of cancer (for a recent review see ref. 2).

TTD is most strikingly characterized by hair abnormalities which are considered as the key factors in the recognition of these patients (Fig. 2). Scalp hair, eyebrows, and eyelashes are short, thin, brittle and dry. Light microscopy reveals irregular hair surface and diameter, trichoschisis, a decreased cuticular layer with twisting, and a nodal appearance that mimicks

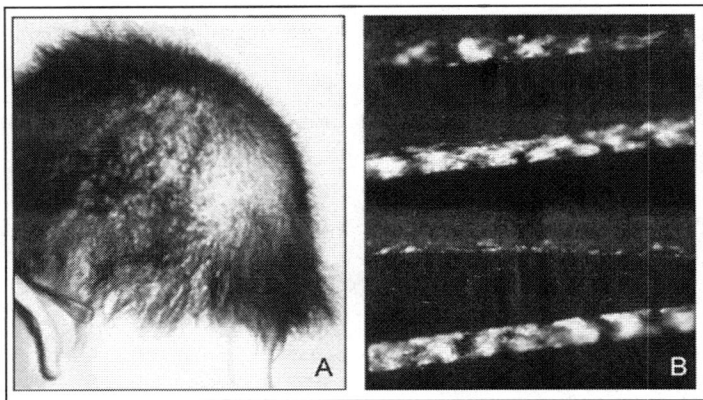

Figure 2. Hair features in TTD (courtesy of Dr. Stefano Marinoni, Trieste, Italy). A) Short, sparse and broken scalp hair. B) Polarizing microscopy of hair shafts showing tiger-tail pattern.

trichorrhexis nodosa. Polarisation microscopy of the hair typically shows alternating light and dark bands that confer a "tiger tail" pattern. Scanning electron microscopy usually reveals absent or severely damaged cuticle scales, an irregular hair surface, trichoschisis and trichorrhexis nodosa, and torsions in the flattened hair shaft. The hair, and often the nails, of affected individuals are characterized by a reduction in cystine, cysteic acid and sulphur content, a reduction that derives from decreases in sulphur-rich matrix proteins. Sulphur proteins are altered not only quantitatively but also qualitatively and there is an abnormal distribution of the sulphur-rich proteins in the cortex and in hair cuticles.

Clinical Photosensitivity Is Usually Associated with an Altered Cellular Response to UV

In 1986, we first reported the results of DNA repair investigations on four Italian patients who showed clinical symptoms diagnostic of TTD together with acute photosensitivity. Upon UV irradiation, cells from these patients showed a notable reduction in levels of survival and UV-induced DNA repair synthesis (UDS), a failure to recover normal DNA and RNA synthesis rate, and an increased mutability.[3] The occurrence of these abnormalities in cellular response to UV light indicates the presence of a defect in nucleotide excision repair (NER), pathway that removes a wide spectrum of DNA lesions, including UV-induced damage. Defects in NER had previously been identified in two other hereditary disorders, namely xeroderma pigmentosum (XP) and Cockayne syndrome (CS), which are characterized by certain unique clinical symptoms. In particular, XP displays various manifestations of cutaneous UV-genotoxicity, notably photosensitivity, pigmentation abnormalities, early onset of precancerous lesions and a greatly increased incidence of cancer in the photoexposed area of the skin. In contrast, cancer-proneness has never been reported in CS patients, in whom skin symptoms are confined to photosensitivity, and the major diagnostic criteria for CS are postnatal growth failure and progressive neurological dysfunction (reviewed in ref. 4).

Although the respective clinical symptoms are distinct, the cellular phenotype that we observed in the four TTD patients was closely similar to that of XP. This finding prompted us to investigate genetic homology between XP and TTD. The investigation consisted of a classical complementation test that was based on the analysis of the capacity of somatic cell hybrids to perform the UDS (Fig. 3). TTD cells were fused with XP cells representative of the seven NER-deficient complementation groups hitherto identified (designated XP-A to XP-G), and

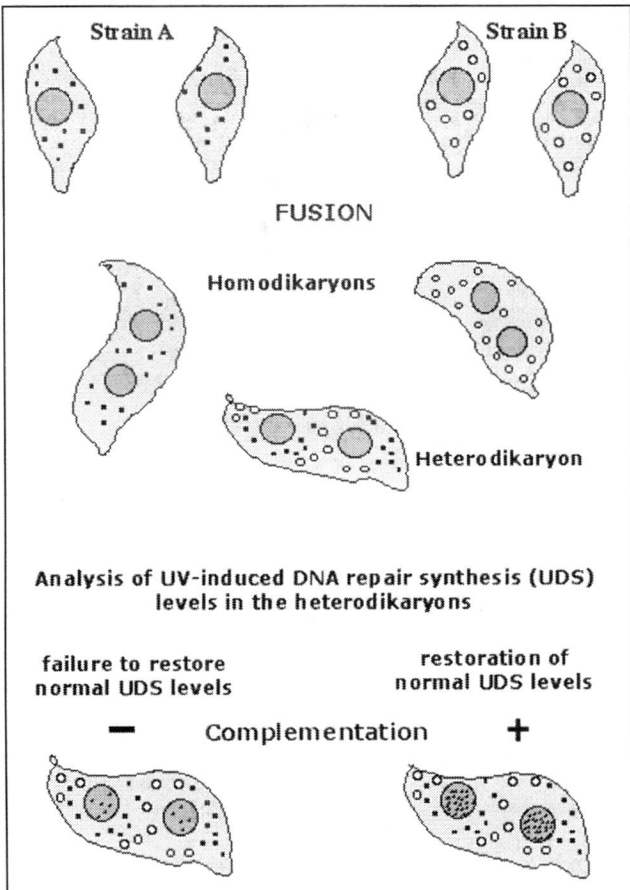

Figure 3. Classical complementation assay by cell fusion for nucleotide excision repair defects. The fibroblast strains used as partners in the fusion are grown for three days in medium containing latex beads of different sizes that are incorporated into the cytoplasm as a marker. The cells are fused using polyethylene glycol and, two days later, analyzed for their ability to perform UV-induced DNA repair synthesis (UDS) by autoradiography. The two cell strains are classified in the same complementation group if the heterodikaryons (identified as binuclear cells containing beads of different sizes) fail to recover normal UDS levels. Conversely, the recovery of normal UDS levels in the heterodikaryons indicates that the cell strains used as partners in the fusion are carrier of genetically different defects.

the UDS level was analyzed in the heterodikaryons. Since parental cells in each cross were labelled with latex beads of two different sizes, heterodikaryons were unambiguously identified as binuclear cells containing beads of both sizes. Restoration of normal UDS levels was observed in all cases except in the crosses between TTD and XP-D cell strains. These results indicated that UV hypersensitivity in the four Italian TTD patients was due to the presence of a defect in XPD.[3]

Since then, DNA repair investigations have been extended to other TTD cases from different countries. The cellular response to UV appeared to be normal in patients with normal cutaneous photosensitivity, and defective in patients who showed photosensitivity, together with other classic diagnostic symptoms for TTD.[5-8]

Heterogeneity of the Repair Defect in TTD

Genetic analysis of the DNA repair defect, performed so far in about forty patients, has led to the identification of three genetically different DNA repair defects associated with TTD. The majority of the repair-deficient TTD patients have the same genetic defect as that present in the XP-D group. This group comprises many patients with XP and two patients who showed clinical symptoms of XP in association with those of CS (for recent reviews see refs. 9,10). In TTD patients mutated in *XPD*, the severity of the pathological phenotype varies from moderate (short stature, delayed puberty, mental development at preschool or primary school level, axial hypotonia, reduced motor coordination and survival beyond early childhood) to severe (very poor mental and motor performances and speech, failure to thrive and death during early childhood). It is of interest that in these patients heterogeneity has been observed also in the degree of severity of the repair defect, with UDS levels ranging between 50% and less than 10% of normal. The Italian patients and some of the non-Italian cases show a drastic reduction in the UDS levels with a cellular sensitivity to UV similar to or even greater than that in XP patients classified in the XP-D group. Other non-Italian TTD patients are characterized by a lower reduction in UDS level with survival significantly affected only at high UV doses.[8] No correlation was observed between the degree of the repair deficiency and the severity of the clinical symptoms.

A different defect was found in one French family with two afflicted children showing a mildly affected phenotype with hair abnormalities but without physical and mental impairment, and partially reduced UDS levels (40% of normal). In these patients the repair defect appeared to be the same as that found in the XP-B group.[11,12] XP-B is a very rare defect and to date has only been identified in two families with the XP/CS phenotype.[13-15]

A new NER defect that has been designated as TTD-A was found in one English patient[16] and, more recently, in three additional cases.[17] TTD-A patients show consistently reduced UDS levels (15-25% of normal) and clinical features of moderate severity. In conclusion, genetic characterization of the repair defect in TTD patients led to the identification of a new gene that is involved in NER (the *TTDA* gene), and to the demonstration that two defects (the XP-B and XP-D defects), which have already been described as responsible for XP and XP/CS pathological phenotypes, are associated with TTD. Therefore, two unexpected findings have emerged from DNA repair investigations on TTD, namely: i) the lack of cutaneous skin abnormalities and skin cancer even in those patients whose DNA repair defect is the same as that of XP-D, and ii) mutations in either *XPB* or *XPD* gene are associated with at least two clinically distinct entities (TTD, XP and/or XP/CS), sharing only cutaneous photosensitivity.

The Genes Mutated in TTD Are All Related to the Transcription Factor TFIIH

These intriguing aspects have been rationalized by the discovery that the genes mutated in TTD encode three distinct subunits of TFIIH, a multiprotein complex that has a dual role in NER and general transcription.[17-20] TTDA is involved in the stabilization of the TFIIH complex,[18] whereas *XPB* and *XPD* encode two subunits of TFIIH.[19,20]

The transcriptionally active form of TFIIH (holo-TFIIH) is made up of a total of ten subunits (Fig. 4). Five of these (XPB, p62, p52, p44 and p34) are tightly associated in a subcomplex called coreTFIIH. XPD is less tightly associated with the core and mediates the binding of the CDK-activating kinase (CAK) subcomplex, which includes the three subunits, cdk7, cyclin H and MAT1 (for reviews see refs. 21-23). The recently identified tenth subunit of TFIIH (TTDA or TFB5) is a 8 KDa protein involved in the stabilization of the complex.[17] XPB and XPD are ATP-dependent helicases with opposite polarity. TFIIH participates in local DNA unwinding, either around the promoter, to allow the synthesis of the

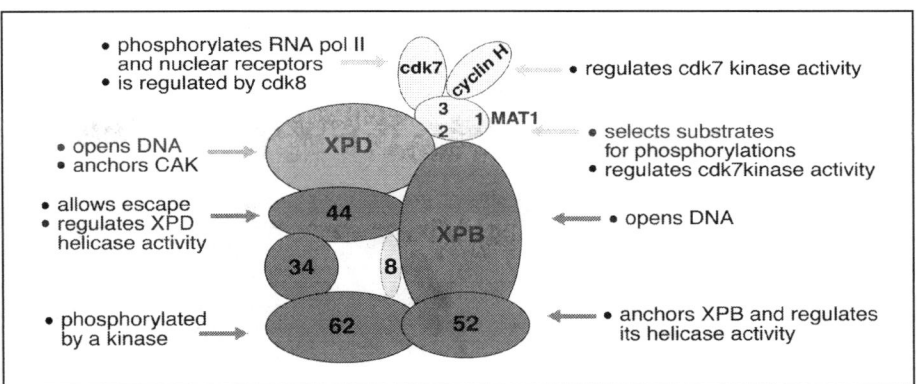

Figure 4. TFIIH composition and enzymatic functions of its subunits. The five subunits composing the coreTFIIH subcomplex are in red, the three subunits of the cdk-activating kinase (CAK) subcomplex are in light blue, the XPD subunit that bridge the two TFIIH subcomplexes is in green and the p8 (TTDA) subunit is in yellow. Adapted from reference 23. A color version of this figure is available online at http://www.Eurekah.com/.

RNA transcript by RNA polymerase II, or around a damaged site, to permit damage-specific nucleases to cleave the DNA on either side of the damage. The 3'-5' helicase activity of XPB is essential for both transcription and repair, whereas the XPD 5'-3' helicase activity is necessary for repair but dispensable for in vitro basal transcription, although XPD substantially stimulates the transcription activity in vitro. This probably accounts for the rarity of XP-B patients compared with the relatively high frequency and variety of pathological phenotypes associated with the XP-D defect. The TFIIH complex also shows ATPase activity associated with the XPD and XPB helicases and kinase activity from the cdk7 subunit, which is able to phosphorylate numerous substrates, including the C-terminal domain (CTD) of the large subunit of RNA polymerase II, converting it from the initiating IIa form to the elongating IIo form. This kinase activity is involved in the regulation of transcription and cell cycle (reviewed in refs. 24,25 respectively), and in the negative control of repair.[26] The enzymatic activities of TFIIH are tightly controlled by interactions within the TFIIH complex or through interactions with many general and regulatory transcription factors. For example, p44 interacts with XPD to stimulate its helicase activity, p52 regulates the function of XPB through pair-wise interactions, MAT1 and cyclinH regulate cdk7 kinase activity, and phosphorylation of cyclinH by cdk8/cyclinC represses both the activation of transcription by TFIIH and the ability of TFIIH to phosphorylate RNA polymerase II. Recently, it has been shown that in *Drosophila* Xpd negatively regulates the CAK activity of Cdk7 and it is down-regulated at the beginning of mitosis, thus contributing to the upregulation of mitotic CAK activity, to the positive regulation of mitotic progression and also, likely, to the mechanism of mitotic silencing of basal transcription.[27] Thus, the structural function of XPD in TFIIH assembly appears to be used by cells in a dynamic way to regulate and coordinate the diverse cellular functions of the different sub-complexes in transcription, DNA repair and cell cycle progression.[25]

The identity of *TTDA*, the third gene responsible for the photosensitive form of TTD, has been for years a key unanswered question. It has been suggested that the function defective in TTD-A is involved in the stabilization of TFIIH or in its protection against degradation.[18] The gene has been recently cloned and its product was shown to have a role in regulating the level of TFIIH.[17]

Table 1. Mutations in the XPB gene found in TTD and XP/CS patients

Clinical Phenotype	Patient Code	Nucleotide Change	Position	Aminoacid Change
TTD mild	TTD4VI[a] TTD6VI[a]	A to C	exon 3	thr119pro
XP/CS mild	XP11BE	C to A[c]	acceptor site of the last intron	Splicing anomaly, frameshift
	XPCS1BA[b] XPCS2BA[b]	T to C[d]	exon 3	phe99ser

[a,b] Brothers; [c] The paternal allele is not expressed; [d] The maternal allele is not expressed.

Mutational Analysis in Repair-Defective TTD Patients

The dual function of TFIIH in repair and transcription led to the notion that clinical features diagnostic for XP result from mutations that interfere only with the DNA repair function of TFIIH, while those typical of TTD and CS are due to a subtle impairment of its transcriptional role.[11,28] This hypothesis requires that the mutations associated with the three disorders are located at different sites in the gene. This was explored by determining the sites of mutations in many patients. Different mutations in the *XPB* gene were found in association with the TTD and the XP/CS phenotype, although the study was limited to five cases from three families, due to the rarity of the XP-B defect (Table 1).[12,15,29] Mutation analysis in XP-D patients has been performed in a significant number of cases, namely twenty nine with XP,[30-35] twenty eight with TTD[32,34,36-39] and two with XP/CS.[31,40] Each mutated site was indeed found in either XP or TTD or XP/CS individuals (Fig. 5). The only exceptions are two alleles, resulting in either substitution of arg616 or a leu461val change associated with deletion of amino acid region 716-730. However, these alleles are completely inactive, at least in *S. pombe* orthologue, suggesting that the phenotype is determined by the second *XPD* allele that was found always different in XP and TTD cases.[32] Most of the mutations are clustered in the C-terminal third of the protein whereas a few are close to the N-terminus and the region between amino acids 260 and 460 is devoid of mutations.[10] The XP phenotype often results from mutations in the helicase motifs of XPD. Although the mutational pattern delineates neither a "TTD domain" nor an "XP domain", a limited number of mutations has consistently been found in association with a given pathological phenotype. Most of the mutations in TTD are localized at three sites (arg112, arg658 and arg722) whereas 70% of mutations in XP patients are localized at a single site, arg683. These may be "TTD-specific" and "XP-specific" mutations. Accordingly, a mouse generated with the TTD-specific arg722trp mutation had many of the features of TTD.[41,42]

Besides providing data supporting the hypothesis that the site of the mutation in the *XPD* gene determines the clinical phenotype, the mutational analysis of the Italian TTD patients has shed new light on the basis of the heterogeneity observed among TTD patients in the degree of severity of the repair defect and of the clinical symptoms. The different degrees of impairment in the cellular responses to UV in TTD appeared to be related to specific mutations.[39] Both in the homozygous patients and in the functionally hemizygous patients, substantial UV sensitivity was associated with the arg112his substitution whereas a mild UV sensitivity was associated with mutations resulting either in the change of arg658 or in the loss of the final portion of the XPD protein. An intermediate UV sensitivity, similar to that found in XP patients defective in *XPD*, was associated with the arg722trp change. In contrast, the severity of the clinical

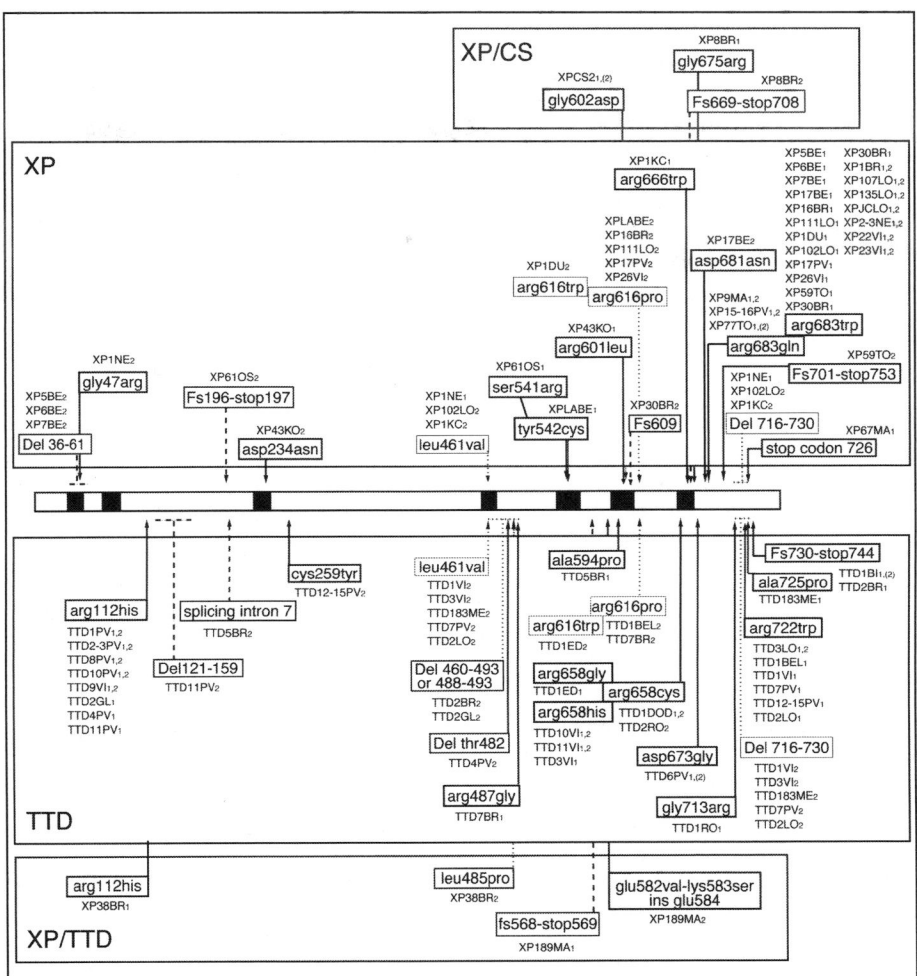

Figure 5. Mutations in the XPD protein in TTD, XP and XP/CS patients. The diagram shows the XPD protein with the helicase domains (black boxes). The amino acid changes resulting from the mutations found in the different pathological phenotypes are shown boxed. The numbers 1 and 2 after the patient code denote the different alleles. The changes responsible for the pathological phenotype, those resulting in deletions likely to affect cellular viability and mutations described as lethal[32, 60] are indicated by solid, dashed and dotted arrows, respectively. The mutation leu461val and the deletion 716-730 have been always found associated in a single haplotype.

symptoms did not correlate with the magnitude of the DNA-repair defect but it appears to be influenced by the dosage of the mutated allele. The most severe clinical features were found in patients who were functionally hemizygous, suggesting that TFIIH in these patients not only contains a mutated XPD subunit, but it also could be present at only half of the normal amount. This may well result in a more severe impairment of the transcriptional activity of TFIIH. This hypothesis implies that, in tissues critically affected in TTD, the level of XPD and, by implication, of TFIIH is rate limiting factor for transcription. Further investigations have demonstrated, however, that a more complex situation underlies the TTD pathological phenotype as well as the clinical outcome of mutations in the *XPD* gene.

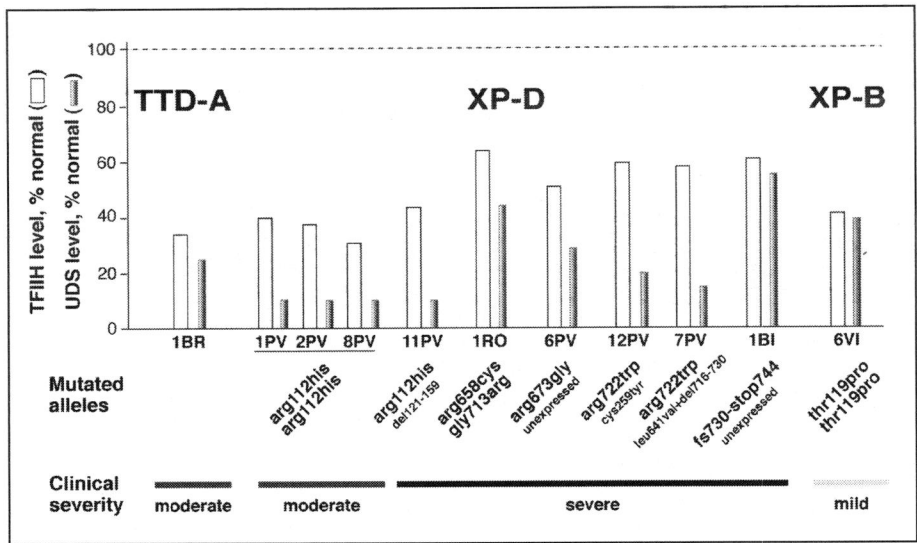

Figure 6. Levels of TFIIH, UV-induced DNA repair synthesis and clinical severity in TTD patients mutated in the *TTDA*, *XPD* and *XPB* gene. The level of TFIIH was analyzed by Western Blot of cell lysates using antibodies against the subunits cdk7, p44 and p62. The reported TFIIH levels refer to the mean steady state of the analyzed subunits in patient fibroblasts expressed as percentage of that in normal fibroblasts analyzed in parallel. UV-induced DNA repair synthesis (UDS) after irradiation with 10 J/m² observed in patient cells is expressed as percentage of that in normal cells. For details see reference 43.

The Cellular Amount of TFIIH Is Reduced in All Repair-Defective TTD Patients

The analysis of the steady-state level of TFIIH in fibroblasts from patients representative of distinct clinical, cellular and molecular alterations demonstrated that mutations in any of the three genes (*XPD*, *XPB* and *TTDA*) responsible for the photosensitive form of TTD cause a decrease by up to 70% in the cellular concentration of TFIIH.[43] The reduction in the amount of TFIIH, however, did not correlate either with the residual repair capacity or with the severity of clinical symptoms (Fig. 6). Substantial reductions in TFIIH levels have been detected in patients showing relatively moderate psychomotor retardation but no increased proneness to infections whereas less marked reductions by 35-45% of normal levels were found in patients who had drastically compromised pathological phenotypes. These observations strikingly indicate that the severity of the TTD pathological phenotype cannot be related solely to the effects of mutations on TFIIH levels.[43]

What is the basis of the reduced amount of TFIIH in TTD cells? In vivo, the ectopic expression of the XPD wild-type protein in *XPD* mutated TTD cells appeared to be sufficient to restore normal levels of the other TFIIH subunits. This increase was paralleled by the restoration of normal repair activity (Fig. 7). These findings indicated that a normal XPD protein provides stability to the TFIIH complex, probably by restoring proper protein-protein interactions, and ensures its correct functioning in NER.[43] In vitro evidence further indicates that a mutated TFIIH subunit may affect the stability of the entire complex. It has been shown that mutations in XPB and p52 prevent the XPB anchoring within the core TFIIH,[44] and mutations in p44 prevent incorporation of the p62 subunit within the core TFIIH.[45] Mutations in XPD or in p44 that modify the XPD-p44 interaction may affect the composition of

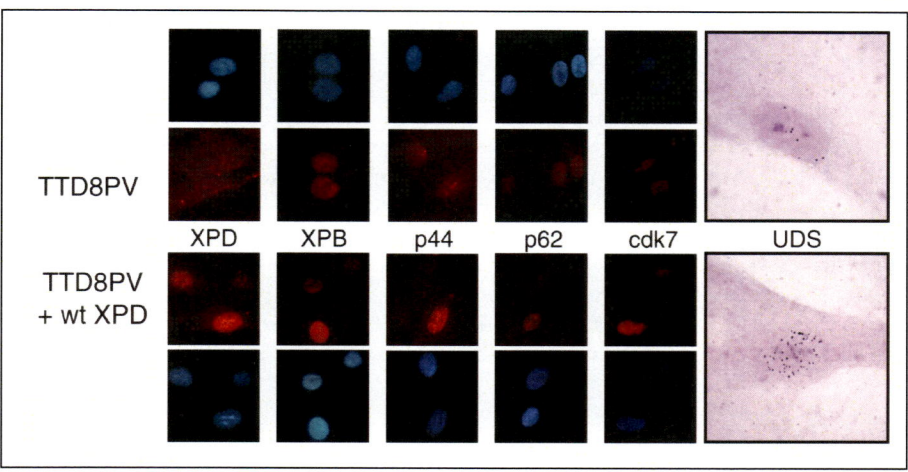

Figure 7. The expression of the XPD wild-type protein in TTD8PV cells restores normal levels of the other TFIIH subunits and normal UV-induced DNA repair synthesis capacity. TTD8PV fibroblasts and stably transfected TTD8PV fibroblasts ectopically expressing the wild-type XPD (TTD8PV+wt XPD) were analyzed for the level of the TFIIH subunits cdk7, p44, p62, XPD and XPB by immunofluorescence (red). Cells were counterstained with Hoechst 33258 (blue). UV-induced DNA repair synthesis capacity (UDS) was analyzed at the single cell level on autoradiographic preparations following irradiation with an UV dose of 20 J/m². For details see reference 43.

TFIIH by decreasing the amount of XPD and CAK subunits associated with the core and/or weakening the anchoring of CAK to the core TFIIH.[46,47] Furthermore, mutations in the C-terminal region of XPD have been shown to play an important role in maintaining TFIIH architecture whereas mutations in the N-terminal region of XPD do not affect either the composition of recombinant TFIIH complexes or the contact with MAT1, the interacting partner of XPD within the CAK complex.[48]

Mutations in the *XPB* and *XPD* genes may render either the transcript or the protein unstable, or interfere with correct folding or proper associations of the protein with the other components of TFIIH, leading to rapid degradation of uncomplexed proteins. Alternatively, a mutated subunit may induce slight conformational changes in the architecture of the TFIIH complex that, in turn, may favor its degradation. This hypothesis may hold true also for the TTD-A defect. Alterations in a factor with a role in the stabilization of the complex might prevent the conformational changes required for optimal functioning of TFIIH. The findings in TTD-A cells suggest that a 3-4 fold reduced amount of TFIIH complexes with normal composition consistently reduce NER efficiency (to 25% of normal) and confers subtle defects in transcription resulting in a TTD phenotype with a physical and mental impairment of moderate severity.[18] Less severe reductions may be necessary but they are not sufficient to generate the TTD phenotype. Reduced levels of TFIIH were indeed found also in some XP-D cell strains from XP and XP/CS patients, although the levels were in general not as low as in the TTD cell strains.[43] This finding already suggested that the clinical outcome of *XPD* mutations is the combined result of the reduction in TFIIH content and the effects of the specific mutations on the interactions of TFIIH with other components of the transcription machinery that may differentially compromise transcription activity. Interestingly, it has been reported that in *Drosophila* having mutations in haywire, the *XPB* homolog, the penetrance of diverse phenotypes also does not correlate with the type of mutation.[49]

Consequences of *XPD* Mutations on the Activities of TFIIH

Explanations for the nature of the various clinical features associated with the XP-D defect have been provided by recent in vitro studies with recombinant TFIIH complexes in which the XPD subunit carries amino acid changes found in patients. All the analyzed mutations, independent of the associated pathological phenotype, appeared to affect the helicase activity of XPD but only those responsible for TTD diminish the basal transcription activity of TFIIH.[48] Inhibition of the XPD helicase activity caused by every mutation found in patients explains the NER deficiency present in all the patients mutated in *XPD*. In particular, the mutations resulting in amino acid changes in the first two-thirds of the protein, including residue 602, abolish the XPD intrinsic helicase activity whereas the mutations in the last C-terminal third of XPD prevent its interaction with p44 and, by consequence, stimulation of the helicase activity of XPD. Remarkably, while mutations associated with TTD confer also a significant in vitro basal transcription defect, mutations associated with the XP or the XP/CS phenotypes, even those that completely disrupt XPD helicase activity, still allow RNA synthesis. Two alleles, resulting in either the substitution of arg616 or the leu461val change associated with the deletion of aminoacid region 716-730, which were found to be null alleles in *S. pombe*,[32] completely abolished basal transcription. A total defect in transcription is obviously incompatible with life. This explains why these two alleles, despite being relatively frequent among patients, have never been found in the homozygous state but always in association with other mutated *XPD* alleles, all of which interfere with basal transcription less drastically.

TTD Symptoms Reflect Subtle Defects in Transcription

Therefore, two specific defects affect the TFIIH complex in TTD cells: a reduced steady-state level and an impaired in vitro transcription activity. A general implication of the reduced amount of TFIIH found in TTD is that a limited availability of TFIIH interferes with NER but is compatible with life. This implies that the TFIIH level albeit reduced in TTD cells does not seriously impede transcription of most genes but it might impair transcription only under certain conditions or in specific cellular compartments affecting, for example, only a limited set of genes that critically demand optimal TFIIH function. Several lines of evidence support the hypothesis that the reduced amount of TFIIH in TTD patients may become limiting in terminally differentiated tissues, in which the mutated TFIIH might get exhausted before the transcriptional program has been completed. The link between TFIIH instability and the hair and skin features that are the diagnostic hallmark of TTD is supported by the peculiar situation described in four TTD patients with fever-dependent reversible deterioration of TTD features.[50] The *XPD* mutation responsible for the pathological phenotype in these patients (resulting in the arg658cys change) confers a temperature-sensitive defect in transcription and repair due to thermo-instability of TFIIH. During the terminal differentiation of hair, nails and skin, the gene family of cysteine-rich matrix proteins is the last to be transcribed at a very high rate: the encoded proteins constitute a major fraction of hair and ensure keratin-crosslinking before cell death. In the case of destabilizing TTD mutations, TFIIH may be depleted before terminal differentiation is complete. As a consequence, crosslinking of keratin filaments is not finished, leading to incomplete (brittle) hair. This phenomenon becomes more pronounced in TTD patients carrying the arg658cys mutation, when high fever further destabilizes TFIIH.[50]

Further evidence supports the hypothesis that many of the clinical features of TTD result from inadequate expression of a diverse set of highly expressed genes. A systematic study of eleven TTD patients demonstrated that all of them showed wholesome hallmarks of β-thalassemia trait: β-globin mRNA levels were reduced in the reticulocytes, and β-globin synthesis was also reduced relative to α-globin. Besides offering an easily measurable diagnostic test for TTD, these data provide a direct evidence for a transcriptional deficiency in TTD.[34] A marked reduction in T cell proliferation in response to mitogens was detected in TTD lymphocytes from TTD patients[3] and alterations in T cells and dendritic cells (DC) suggestive of a subtle transcriptional defect of a set of genes involved in DC maturation and function, have been reported in a TTD child with a severe immunodeficiency.[51] These observations are paralleled by the finding of reduced transcription of the skin-specific, differentiation-related gene *SPRR2* in the TTD mouse expressing the arg722trp mutated XPD protein.[41] The SPRR2 gene is a member of the small proline-rich protein (SPRR) family expressed in epidermis. It encodes a structural component of the cornified envelope and is expressed in the final stage of terminal differentiation. Reduced SPRR2 expression in TTD mouse skin reflects defective gene transcription in late stages of terminally differentiating epidermal keratinocytes.

The observation that *XPD* mutations associated with TTD interfere with transcription may help to explain the lack of cancer proneness in TTD, despite the reduced efficiency of NER. The defect in transcription together with limiting amounts of TFIIH could prevent tumoral transformation and/or progression. However, also the involvement of TFIIH in cell-cycle regulation could play a role in the lack of cancer susceptibility in TTD. All the studies performed so far, have not identified any cellular and biochemical parameter whose alteration unequivocally correlates with the different cancer susceptibility in XP and TTD patients defective in the same repair pathway, and even in the same gene (reviewed in ref. 9).

It has been suggested that the other symptoms of TTD, namely those that confer the aging phenotype, are caused by unrepaired oxidative damage that compromises transcription and leads to functional inactivation of critical genes and enhanced apoptosis, resulting ultimately in functional decline and depletion of cell renewal capacity.[42] In this context, it is worthwhile mentioning that defects in RNA pol I transcription have been recently suggested as an alternative explanation for the pathological phenotype of CS, a disorder that shares many features of aging and developmental anomalies with TTD.[52] The involvement of TFIIH in transcription by RNA polymerase I has been demonstrated both in vitro[53] and in vivo.[54] Therefore, it cannot be ruled out that a mutated XPB or XPD subunit might interfere with the role of TFIIH in RNA pol I transcription, thus accounting for the overlapping symptoms in CS and TTD.

XPD Mutations May Interfere with the Regulation of Gene Expression by TFIIH

An extra layer of complexity in the interpretations of mutational effects of TFIIH has been added by recent discoveries that extend the roles of TFIIH in transcription (reviewed in ref. 24). Apart from its role in basal transcription by RNA polymerase I and II, TFIIH maintains complicated cross-talk with different types of factors involved in RNA Pol II-mediated transcription and has a regulatory role during transcription.

It has been shown that TFIIH interacts with some nuclear hormone receptors, including the estrogen (ERα), androgen (AR) and retinoic acid (RARα and RARγ) receptors (reviewed in ref. 55). As a consequence of their interaction with TFIIH, ERα and RARs are phosphorylated in their N-terminal A/B region by cdk7, a component of the TFIIH CAK subcomplex, which is anchored to the coreTFIIH by XPD. This phosphorylation process plays a critical role in transcription mediated by the liganded receptors, allowing ligand-dependent control of the activation of the hormone-responsive genes. A reduced ligand-dependent activation of transcription by the three receptors RARα, ERα and AR was observed in XP-D cells with the mutation arg683trp, i.e., the change most frequently found in XP patients.[56] Conversely, the arg683trp mutation had no effect on the transactivation mediated by the Vitamin D receptor, which lacks the typical A/B domain targeted by cdk7.[48] Transcriptional activation by RARα was significantly reduced also in one TTD cell line with the arg722trp change whereas almost optimal transactivation was found in a TTD cell line with the arg112his substitution and in TTD-A cells.[48,56] These findings indicate that mutations located in the C-terminal end of XPD, independent of the associated pathological phenotype, prevent the action of nuclear receptors having an A/B domain. This might account for the developmental and neurological defects encountered in both XP and TTD patients having mutations at position arg658 (TTD), arg683trp (XP), and arg722trp (TTD).[48]

TFIIH interacts also with some tissue specific transactivators. It has been demonstrated that the transcriptional activator FUSE Binding protein (FBP), a regulator of *c-myc* expression, binds specifically to TFIIH.[57] In two cell lines defective in either the *XPB* or *XPD* gene (derived from the XP/CS patient XP11BE and from the XP patient XP6BE, respectively), this interaction was either abolished or attenuated, resulting in impaired regulation of *c-myc* expression.[58] Mutations altering TFIIH conformation may therefore affect to different degrees its stereo-specific interactions with tissue-specific transcription regulators (activators and repressors). We can speculate that while XP-type mutations could interfere with a specific class of regulators involved in carcinogenesis, TTD-type mutations may affect the interaction of TFIIH with transcription factors involved in other regulated pathways such as differentiation, development and neurogenesis. The possibility that TTD mutations differentially affect activated transcription mediated by the interaction of TFIIH with different transcription regulators may be an alternative explanation of the lack of cancer proneness in TTD, despite the reduced efficiency of NER.

TFIIH Defects and Clinical Outcome

It is apparent from recent studies that the activity of TFIIH in transcription is multifaceted, ranging from transcription by RNA pol I and RNA pol II to regulation of gene expression. It has also become clear that the role of TFIIH in fundamental cellular processes, transcription, DNA repair and cell-cycle, is more complex than initially thought, which has important implications for the interpretation of phenotypes caused by alterations in TFIIH. The effect of mutations in the *XPB* or *XPD* genes may each be subtly different, affecting in slightly different ways the stability and the conformation of TFIIH and, consequently, its functional activities. The phenotypic consequences of a mutated allele will depend on the precise balance and inter-relationships between these effects and the clinical outcome in patients will reflect the combined effects of each allele on TFIIH activity and the gene dosage. This might explain the puzzling variety of pathological phenotypes, ranging from Cerebro-Oculo-Facio-Skeletal syndrome (COFS)[59] to combined XP/TTD features,[60] that have been recently identified in association with defects in XPD. A complex phenotype of moderate severity with some features of both XP and TTD has been found in two patients, coded XP38BR and XP189MA.[60] In both patients, polarized light microscopy revealed a tiger-tail appearance of the hair, and amino acid analysis of the hairshafts showed levels of sulfur-containing proteins between those of normal and TTD individuals. The patient XP189MA, showing undetectable levels of UDS but mild features of XP, was compound heterozygote for two novel mutations, one resulting in a protein truncation likely to be totally nonfunctional (Fig. 5). It might well be that the mutation present in the less severely affected allele confers an extremely mild defect in transcription that does not completely prevent the phenotypic consequences of the repair defect, as usually found for the mutations associated with TTD. Alternatively, a peculiar situation in the genetic background of this patient may mitigate the defects in both transcription and NER, resulting in mild TTD and XP features.

In the patient XP38BR one of the *XPD* alleles contained a novel mutation resulting in the leu485pro change, which appeared to be lethal in *S. pombe* (Fig. 5). The second allele contained the mutation G413A, causing the arg112his substitution, that has been found in several patients, all with the clinical features of TTD, yet the major features of XP38BR were of XP. The finding of a single mutation resulting in more than one phenotype is unprecedented for NER-defective syndrome, with only one exception. The same set of mutated *CSB* alleles has been observed in two brothers with a severe form of XP and in one patient with the classical form of CS.[61] A further unexpected finding in XP38BR was that the UDS level was substantially higher than that in patients with the same mutation but the TTD phenotype (30% compared to 10% of normal). The most likely explanation for these perplexing findings is that this patient contains an as yet unidentified modifying mutation in another gene that partially suppresses the defects in both transcription and NER that are usually associated with arg112his alteration. As well as resulting in only partial TTD features, this milder putative transcription defect might permit the development of the skin abnormalities characteristic of XP, albeit again with a mild phenotype.

It is worthwhile mentioning that we have found the arg112his change in the state of functional hemizygosity in an Italian patient that shows a severe physical and mental impairment without the hair abnormalities typical of TTD. No cellular and biochemical alteration has been so far identified that may account for the unexpected genotype-phenotype relation in this patient (our unpublished observations). Thus, unusual phenotypes related to defects in the *XPD* gene are emerging and perhaps this is not so surprising in view of the complexity of the role of the TFIIH complex and of its multifaceted activities. During transcription, TFIIH interacts with a variety of factors, including tissue-specific transcription factors, nuclear receptors, chromatin remodeling complexes and RNA, suggesting that in some cases the genetic background may also play a role in the clinical outcome. In this context, the variety of clinical features associated with XP-D defects provides a unique tool to dissect the complex interplay between repair and transcription and the phenotypic consequences of mutations that affect the stability and/or the activity in repair and transcription of the TFIIH complex.

The finding that TFIIH is central to the onset of the photosensitive form of TTD, suggests that the search for the genes implicated in the nonphotosensitive form should be addressed to genes encoding factors of transcriptional regulation. This is still an unexplored research field.

In conclusion, although it is evident that there is still much more to learn about TTD, the last few years have witnessed significant progress in our understanding of the genetic and molecular bases of this disorder. We are also beginning to understand the links between molecular defects and clinical symptoms in TTD: the pathological phenotype that was difficult to explain on the basis of a repair defect appeared to be correlated to subtle defects in transcription. We have still to fully understand why a defect in the same gene causes either a cancer-prone phenotype as in XP, or multi-system abnormalities, as in TTD and how the transcriptional deficiency results in a cancer-free phenotype, even when NER is severely reduced. In this perspective, it is likely that future research will help in elucidating not only the biological roles of the functions defective in TTD, but also the crucial aspects of tumor formation, aging, differentiation and development.

Acknowledgements

I am very grateful to Alan Lehmann for our long-lasting and fruitful collaboration characterised by stimulating discussions and pleasant interactions. I acknowledge Tania Pedrini and Jean-Marc Egly for lively and helpful discussions. I thank Elena Botta, Tiziana Nardo and all the other members of my group at the IGM CNR, Pavia for their contribution to the work over the years. Our studies mentioned in the text have been supported by grants from the Associazione Italiana per la Ricerca sul Cancro (AIRC), the EC (grants SC1-232, CHRX-CT94-0443 and QLG1-1999-00181) and the MIUR (Functional Genomics Project and FIRB grant RBNE01RNN7).

References

1. Price VH, Odom RB, Ward WH et al. Trichothiodystrophy: Sulfur-deficient brittle hair as a marker for a neuroectodermal symptom complex. Arch Dermatol 1980; 116:1375-84.
2. Itin PH, Sarasin A, Pittelkow MR. Trichothiodystrophy: Update on the sulfur-deficient brittle hair syndromes. J Am Acad Dermatol 2001; 44:891-920; quiz 921-4.
3. Stefanini M, Lagomarsini P, Arlett CF et al. Xeroderma pigmentosum (complementation group D) mutation is present in patients affected by trichothiodystrophy with photosensitivity. Hum Genet 1986; 74:107-12.
4. Lehmann AR. DNA repair-deficient diseases, xeroderma pigmentosum, Cockayne Syndrome and trichothiodystrophy. Biochimie; in press.
5. Stefanini M, Lagomarsini P, Giorgi R et al. Complementation studies in cells from patients affected by trichothiodystrophy with normal or enhanced UV photosensitivity. Mutat Res 1987; 191:117-9.
6. Lehmann AR, Arlett CF, Broughton BC et al. Trichothiodystrophy, a human DNA repair disorder with heterogeneity in the cellular response to ultraviolet light. Cancer Res 1988; 48:6090-6.
7. Stefanini M, Giliani S, Nardo T et al. DNA repair investigations in nine Italian patients affected by trichothiodystrophy. Mutat Res 1992; 273:119-25.
8. Stefanini M, Lagomarsini P, Giliani S et al. Genetic heterogeneity of the excision repair defect associated with trichothiodystrophy. Carcinogenesis 1993; 14:1101-5.
9. Bergmann E, Egly JM. Trichothiodystrophy, a transcription syndrome. Trends Genet 2001; 17:279-86.
10. Lehmann AR. The xeroderma pigmentosum group D (XPD) gene: One gene, two functions, three diseases. Genes Dev 2001; 15:15-23.
11. Vermeulen W, van Vuuren AJ, Chipoulet M et al. Three unusual repair deficiencies associated with transcription factor BTF2(TFIIH): Evidence for the existence of a transcription syndrome. Cold Spring Harb Symp Quant Biol 1994; 59:317-29.
12. Weeda G, Eveno E, Donker I et al. A mutation in the XPB/ERCC3 DNA repair transcription gene, associated with trichothiodystrophy. Am J Hum Genet 1997; 60:320-9.
13. Robbins JH, Kraemer KH, Lutzner MA et al. Xeroderma pigmentosum. An inherited diseases with sun sensitivity, multiple cutaneous neoplasms, and abnormal DNA repair. Ann Intern Med 1974; 80:221-48.
14. Scott RJ, Itin P, Kleijer WJ et al. Xeroderma pigmentosum-cockayne syndrome complex in two patients: Absence of skin tumors despite severe deficiency of DNA excision repair. J Am Acad Dermatol 1993; 29:883-9.
15. Vermeulen W, Scott RJ, Rodgers S et al. Clinical heterogeneity within xeroderma pigmentosum associated with mutations in the DNA repair and transcription gene ERCC3. Am J Hum Genet 1994; 54:191-200.
16. Stefanini M, Vermeulen W, Weeda G et al. A new nucleotide-excision-repair gene associated with the disorder trichothiodystrophy. Am J Hum Genet 1993; 53:817-21.
17. Giglia-Mari G, Coin F, Ranish JA et al. A new, tenth subunit of TFIIH is responsible for the DNA repair syndrome trichothiodystrophy group A and stabilizes TFIIH. Nat Genet 2004; 36:714-9.
18. Vermeulen W, Bergmann E, Auriol J et al. Sublimiting concentration of TFIIH transcription/DNA repair factor causes TTD-A trichothiodystrophy disorder. Nat Genet 2000; 26:307-13.
19. Schaeffer L, Roy R, Humbert S et al. DNA repair helicase: A component of BTF2 (TFIIH) basic transcription factor. Science 1993; 260:58-63.
20. Schaeffer L, Moncollin V, Roy R et al. The ERCC2/DNA repair protein is associated with the class II BTF2/TFIIH transcription factor. EMBO J 1994; 13:2388-92.
21. Coin F, Egly JM. Ten years of TFIIH. Cold Spring Harb Symp Quant Biol 1998; 63:105-10.
22. Dvir A, Conaway JW, Conaway RC. Mechanism of transcription initiation and promoter escape by RNA polymerase II. Curr Opin Genet Dev 2001; 11:209-14.
23. Egly JM. The 14th Datta Lecture. TFIIH: From transcription to clinic. FEBS Lett 2001; 498:124-8.
24. Zurita M, Merino C. The transcriptional complexity of the TFIIH complex. Trends Genet 2003; 19:578-84.
25. Chen J, Suter B. Xpd, a structural bridge and a functional link. Cell Cycle 2003; 2:503-6.

26. Araujo SJ, Tirode F, Coin F et al. Nucleotide excision repair of DNA with recombinant human proteins: Definition of the minimal set of factors, active forms of TFIIH, and modulation by CAK. Genes Dev 2000; 14:349-59.

27. Chen J, Larochelle S, Li X et al. Xpd/Ercc2 regulates CAK activity and mitotic progression. Nature 2003; 424:228-32.

28. Bootsma D, Hoeijmakers JH. DNA repair. Engagement with transcription. Nature 1993; 363:114-5.

29. Weeda G, van Ham RC, Vermeulen W et al. A presumed DNA helicase encoded by ERCC-3 is involved in the human repair disorders xeroderma pigmentosum and Cockayne's syndrome. Cell 1990; 62:777-91.

30. Frederick GD, Amirkhan RH, Schultz RA et al. Structural and mutational analysis of the xeroderma pigmentosum group D (XPD) gene. Hum Mol Genet 1994; 3:1783-8.

31. Takayama K, Salazar EP, Lehmann A et al. Defects in the DNA repair and transcription gene ERCC2 in the cancer- prone disorder xeroderma pigmentosum group D. Cancer Res 1995; 55:5656-63.

32. Taylor EM, Broughton BC, Botta E et al. Xeroderma pigmentosum and trichothiodystrophy are associated with different mutations in the XPD (ERCC2) repair/transcription gene. Proc Natl Acad Sci USA 1997; 94:8658-63.

33. Kobayashi T, Kuraoka I, Saijo M et al. Mutations in the XPD gene leading to xeroderma pigmentosum symptoms. Hum Mutat 1997; 9:322-31.

34. Viprakasit V, Gibbons RJ, Broughton BC et al. Mutations in the general transcription factor TFIIH result in beta- thalassaemia in individuals with trichothiodystrophy. Hum Mol Genet 2001; 10:2797-2802.

35. Kobayashi T, Uchiyama M, Fukuro S et al. Mutations in the XPD gene in xeroderma pigmentosum group D cell strains: Confirmation of genotype-phenotype correlation. Am J Med Genet 2002; 110:248-52.

36. Broughton BC, Steingrimsdottir H, Weber CA et al. Mutations in the xeroderma pigmentosum group D DNA repair/transcription gene in patients with trichothiodystrophy. Nat Genet 1994; 7:189-94.

37. Takayama K, Salazar EP, Broughton BC et al. Defects in the DNA repair and transcription gene ERCC2(XPD) in trichothiodystrophy. Am J Hum Genet 1996; 58:263-70.

38. Takayama K, Danks DM, Salazar EP et al. DNA repair characteristics and mutations in the ERCC2 DNA repair and transcription gene in a trichothiodystrophy patient. Hum Mutat 1997; 9:519-25.

39. Botta E, Nardo T, Broughton BC et al. Analysis of mutations in the XPD gene in Italian patients with trichothiodystrophy: Site of mutation correlates with repair deficiency, but gene dosage appears to determine clinical severity. Am J Hum Genet 1998; 63:1036-48.

40. Broughton BC, Thompson AF, Harcourt SA et al. Molecular and cellular analysis of the DNA repair defect in a patient in xeroderma pigmentosum complementation group D who has the clinical features of xeroderma pigmentosum and Cockayne syndrome. Am J Hum Genet 1995; 56:167-74.

41. de Boer J, de Wit J, van Steeg H et al. A mouse model for the basal transcription/DNA repair syndrome trichothiodystrophy. Mol Cell 1998; 1:981-90.

42. de Boer J, Andressoo JO, de Wit J et al. Premature aging in mice deficient in DNA repair and transcription. Science 2002; 296:1276-9.

43. Botta E, Nardo T, Lehmann AR et al. Reduced level of the repair/transcription factor TFIIH in trichothiodystrophy. Hum Mol Genet 2002; 11:2919-28.

44. Jawhari A, Laine JP, Dubaele S et al. p52 mediates XPB function within the transcription/repair factor TFIIH. J Biol Chem 2002; 277:31761-7.

45. Tremeau-Bravard A, Perez C, Egly JM. A role of the C-terminal part of p44 in the promoter escape activity of transcription factor IIH. J Biol Chem 2001; 276:27693-7.

46. Seroz T, Perez C, Bergmann E et al. p44/SSL1, the regulatory subunit of the XPD/RAD3 helicase, plays a crucial role in the transcriptional activity of TFIIH. J Biol Chem 2000; 275:33260-6.

47. Coin F, Bergmann E, Tremeau-Bravard A et al. Mutations in XPB and XPD helicases found in xeroderma pigmentosum patients impair the transcription function of TFIIH. EMBO J 1999; 18:1357-66.

48. Dubaele S, Proietti De Santis L, Bienstock RJ et al. Basal transcription defect discriminates between xeroderma pigmentosum and trichothiodystrophy in XPD patients. Mol Cell 2003; 11:1635-46.

49. Merino C, Reynaud E, Vazquez M et al. DNA repair and transcriptional effects of mutations in TFIIH in Drosophila development. Mol Biol Cell 2002; 13:3246-56.

50. Vermeulen W, Rademakers S, Jaspers NG et al. A temperaturesensitive disorder in basal transcription and DNA repair in humans. Nat Genet 2001; 27:299-303.

51. Racioppi L, Cancrini C, Romiti ML et al. Defective dendritic cell maturation in a child with nucleotide excision repair deficiency and CD4 lymphopenia. Clin Exp Immunol 2001; 126:511-518.

52. Bradsher J, Auriol J, Proietti de Santis L et al. CSB is a component of RNA pol I transcription. Mol Cell 2002; 10:819-29.

53. Iben S, Tschochner H, Bier M et al. TFIIH plays an essential role in RNA polymerase I transcription. Cell 2002; 109:297-306.

54. Hoogstraten D, Nigg AL, Heath H et al. Rapid switching of TFIIH between RNA polymerase I and II transcription and DNA repair in vivo. Mol Cell 2002; 10:1163-74.

55. Rochette-Egly C. Nuclear receptors: Integration of multiple signalling pathways through phosphorylation. Cell Signal 2003; 15:355-66.

56. Keriel A, Stary A, Sarasin A et al. XPD mutations prevent TFIIH-dependent transactivation by nuclear receptors and phosphorylation of RARalpha. Cell 2002; 109:125-35.

57. Liu J, He L, Collins I et al. The FBP interacting repressor targets TFIIH to inhibit activated transcription. Mol Cell 2000; 5:331-41.

58. Liu J, Akoulitchev S, Weber A et al. Defective interplay of activators and repressors with TFIH in xeroderma pigmentosum. Cell 2001; 104:353-63.

59. Graham Jr JM, Anyane-Yeboa K, Raams A et al. Cerebro-oculo-facio-skeletal syndrome with a nucleotide excision-repair defect and a mutated XPD gene, with prenatal diagnosis in a triplet pregnancy. Am J Hum Genet 2001; 69:291-300.

60. Broughton BC, Berneburg M, Fawcett H et al. Two individuals with features of both xeroderma pigmentosum and trichothiodystrophy highlight the complexity of the clinical outcomes of mutations in the XPD gene. Hum Mol Genet 2001; 10:2539-2547.

61. Colella S, Nardo T, Botta E et al. Identical mutations in the CSB gene associated with either cockayne syndrome or the DeSanctis-cacchione variant of xeroderma pigmentosum. Hum Mol Genet 2000; 9:1171-5.

CHAPTER 4

Roles of the BRCA1 and BRCA2 Breast Cancer Susceptibility Proteins in DNA Repair

Katrin Gudmundsdottir, Emily Witt and Alan Ashworth

Abstract

Since the cloning of the *BRCA1* and *BRCA2* genes less than 10 years ago, a great deal of effort has been expended in attempting to uncover the functions of the encoded proteins. BRCA1 and BRCA2 have now been linked to a wide variety of cellular functions through binding or colocalization with other proteins. In this chapter we focus on the best characterized of these potential functions, DNA repair.

The *BRCA* Genes

Both *BRCA1* and *BRCA2* are believed to function as tumour suppressor genes since the wild type allele is frequently lost in the tumors of heterozygous carriers.[1,2] Between 5-10% of all breast cancers are believed to be caused by hereditary predisposition and a significant proportion of these are due to mutations in either *BRCA1* or *BRCA2*. The more breast cancer cases there are in a family and the younger they are at the time of diagnosis the more likely it is that they carry a *BRCA1* or *BRCA2* mutation. Carriers of a *BRCA1* mutation also have a high risk of ovarian cancer, accounting for the majority of breast-ovarian cancer families. *BRCA2* carriers have a significant but lower risk of ovarian cancer, however, they account for most of the male and female breast cancer families.[3] Increased risk of cancer at other sites, such as cancer of the prostate and colon, has also been observed.[4,5] The degree of breast cancer risk conferred by germ line mutations in the *BRCA* genes has been reported to vary between 36% and 85% in females by the age of 70. The risk is generally lower for *BRCA2* mutation carriers than for carriers of a *BRCA1* mutation.[6] The reason for this variance in the penetrance of *BRCA* mutations in individuals is not fully clear, but they have been suggested to involve both the type and position of the mutation in the *BRCA* gene as well as modifying genes and environmental factors.[6]

Somatic *BRCA* mutations have rarely been detected in sporadic cancer cases. Nevertheless, reduced expression of these genes has been observed in sporadic breast carcinomas. In some of these tumors *BRCA1* is silenced by methylation[7] or by gross chromosomal rearrangements, which are thought to be due to *BRCA1* being situated at a chromosomal region rich in Alu repeats.[8] Hypermethylation of the *BRCA2* promoter has been reported in one ovarian tumour[9] but not so far in breast carcinomas.[10]

DNA Repair and Human Disease, edited by Adayabalam S. Balajee. ©2006 Landes Bioscience and Springer Science+Business Media.

BRCA2-related breast cancers do not generally appear to differ significantly from sporadic cancers in terms of pathological features but do tend to be of a higher overall grade. However, breast cancers in patients with a *BRCA1* mutation tend not just to be of a higher grade than sporadic but also have higher mitotic counts, a greater degree of nuclear pleomorphism, and less tubule formation and are usually estrogen receptor negative. *BRCA1* carriers also have an excess of medullary and atypical medullary cancers.[11]

BRCA Genes as Caretakers of the Genome

The *BRCA1* gene was identified in 1994,[12] having been mapped to chromosome 17q by genetic linkage analysis in 1990.[13] It is a large gene of 24 exons, 22 of which are translated into a protein of 1863 amino acids. BRCA1 contains an N-terminal Ring finger domain and two C-terminal BRCT (BRCA1-C-terminal) domains, which are both involved in protein-protein interactions (Fig. 1). The N-terminal Ring finger domain interacts with BARD1 (BRCA1-associated RING domain protein) to form a RING-finger heterodimeric complex that has E3 ubiquitin ligase activity.[14,15] BRCT domains have been found in proteins involved in DNA repair and were recently reported to be a phosphopeptide binding motif.[16,17] BRCA1 has been shown to colocalize and bind to a variety of proteins involved in various cellular processes, such as DNA damage response, DNA repair, transcription, chromatin remodeling and protein ubiquitination.[18]

The *BRCA2* gene was localized to chromosome 13q in 1994[19] and identified a year later.[20] *BRCA2* consists of 27 exons and encodes a large protein of 3418 amino acids, which has little sequence similarity to other proteins to indicate its function. The gene is characterized by a very large central exon 11, which contains 8 copies of a 30-40 amino acid repeat known as the BRC repeats (Fig. 1).[21] It is via these, and also through an unrelated domain at the C-terminus, that BRCA2 binds to the essential DNA recombination and repair protein, RAD51.[22,23]

Both BRCA1 and BRCA2 show similar expression patterns and sub-cellular localization; they are both nuclear proteins, preferentially expressed during the late G1/early-S phase of the cell cycle in a wide range of tissue cell types.[24-26] In mice, the highest expression levels are observed in proliferating and differentiating epithelial tissue, which, in the mammary gland, occurs during puberty, pregnancy and lactation.[27,28]

A number of different mutations introduced into the *Brca* genes in the mouse revealed that they are important for normal embryogenesis.[29] Null mutations for either gene results in early embryonic lethality due to failure of proliferation, accompanied by an elevated expression of the cell cycle inhibitor p21 and p53.[30,31] Consistent with this, the embryonic lethality may be modulated by mutations in the *p53* gene, allowing the embryos to survive a day or two longer.[30,32]

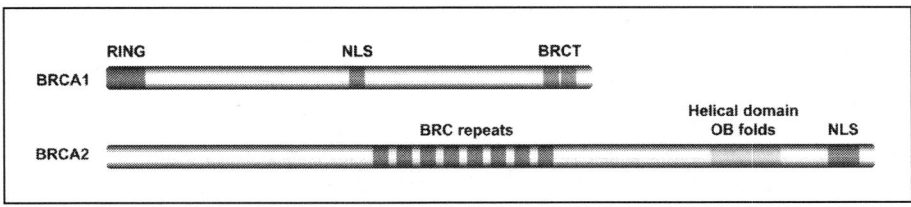

Figure 1. Functional domains of BRCA1 and BRCA2. BRCA1 is characterized by a N-terminal RING finger domain, which has ubiquitin ligase activity, and a C-terminal BRCT domain, which is found in proteins involved in DNA repair. BRCA2 contains eight BRC repeats that bind RAD51 and at the C-terminus is a domain with α-helices and oligonucleotide/oligosaccharide-binding (OB) folds that have been shown to bind DNA and DSS1. Both proteins cotain nuclear localization sequences (NLS).

The p53-dependent growth arrest associated with deficiency of either *BRCA* gene is thought to represent a checkpoint response due to unrepaired chromosome breaks. Indeed, BRCA1 and BRCA2 deficient cells exhibit a high degree of spontaneous and induced chromosome aberrations and they also show hypersensitivity to DNA cross-linking agents, such as mitomycin C, and a low to moderate sensitivity to ionizing radiation (IR).[33-37] This suggests that BRCA1 and BRCA2 deficient cells have a defect in the repair of double strand breaks.

Double Strand Break Repair

The accurate repair of damaged DNA is essential for the maintenance of genomic integrity. There are a number of DNA repair mechanisms, which have evolved to repair DNA following damage. Lesions affecting only one of the DNA strands, in which case the intact complementary strand can be used as a template for repair, are repaired by base-excision repair (BER), nucleotide-excision repair (NER) or mismatch repair (MMR). Although partly overlapping, these pathways show a preference for specific DNA lesions. BER targets small chemical alterations of bases, while NER preferentially repairs helix-distorting lesions and MMR corrects mispaired bases.[38] DNA double strand breaks (DSBs), however, are more problematic since the complementary strand is not available as a template for repair. They are extremely toxic to cells and can be lethal if not repaired. DSBs may come about as a result of either exogenous insults, such as exposure to IR or DNA cross linking drugs, or endogenous events, such as the collapse of replication forks upon encountering a single-strand break. Two major pathways exist in mammalian cells for the repair of DNA DSBs. These are known as Non Homologous End Joining (NHEJ) and Homologous Recombination (HR) (Fig. 2).[39-41] A fundamental difference between these two pathways lies in the requirement of a homologous DNA template for repair by HR while NHEJ does not rely on sequence homology. Therefore, NHEJ tends to be a more error-prone repair pathway.

NHEJ involves the direct religation of broken termini without use of the sister chromatid as a template. Initially, Ku70 and Ku80 form a heterodimer that binds the termini of the DSB. DNA-PK$_{cs}$ is then recruited to form a complex with the Ku heterodimer. Religation of the ends of the DSB is then carried out by DNA ligase IV and XRCC4. NHEJ often results in insertions or deletions of nucleotides at the repair site, leading to mutations within the genome.[39-41]

Homology directed repair of DSBs can be subdivided into two repair pathways, gene conversion (GC) and single-strand annealing (SSA).[42] GC is a RAD51-dependent DNA repair pathway that utilizes either the identical sister chromatid or the homologous chromosome to repair a DSB.[43] It involves the processing of the ends of the DSB to leave 3' overhangs to which RAD52 and RPA (replication protein A) can bind. RAD51 is then recruited to the break site, polymerizing onto the ssDNA to form a nucleoprotein filament. RAD51 catalyzes the invasion of the homologous sequence on the sister chromatid or homologous chromosome, which is then used as a template for accurately repairing the broken ends by DNA synthesis. Recombination intermediates are then resolved either with or without sequence exchange. This pathway is generally thought to be error-free and most active in the S/G2 phases of the cell cycle following DNA replication.[38,43]

SSA is a RAD51-independent DNA repair pathway that also requires a homologous sequence for repair. As with GC, single strand overhangs are created onto which RAD52 binds. Regions of homology on either site of the DSB are then aligned and annealed resulting in the deletion of the intervening sequence. Like NHEJ, this mechanism of DNA repair is error-prone, and may result in large deletions and translocations.[44]

Both *BRCA1* and *BRCA2* appear to be involved in the accurate repair of DNA DSBs by HR and their potential roles within this pathway will be discussed further below.

Figure 2. Mechanism of double strand break repair. Two main pathways exist in mammalian cells for DSB repair. Non-homologous end joining (NHEJ) does not rely on sequence homology when repairing DSBs, instead, broken termini are bound by the Ku heterodimer/DNA-PKcs complex and the ends are simply religated by DNA ligase IV and XRCC4. The homology-directed repair pathway uses a homologous DNA template for repair of DSBs and involves the nucleolytic processing of the DNA ends onto which RPA and RAD52 bind. In gene conversion (GC), RAD51 forms a nucleoprotein filament on the ssDNA and searches for a homologous sequence, which it invades and DNA polymerase uses as a template for repair. Ligation and resolution of recombination intermediates results in accurate repair of the DSB. The orchestration of the RAD51 response may also require RAD51 paralogues and BRCA1 and BRCA2. Single-strand annealing (SSA) is a RAD51-independent pathway that also requires a homologous sequence for the repair of DSBs. This pathway involves the creation of long single-strand overhangs, that are annealed once a homology is exposed, followed by trimming of the intermediate sequence by the ERCC1/XPF nuclease. Both NHEJ and SSA are thought to be more error-prone repair pathways than GC.

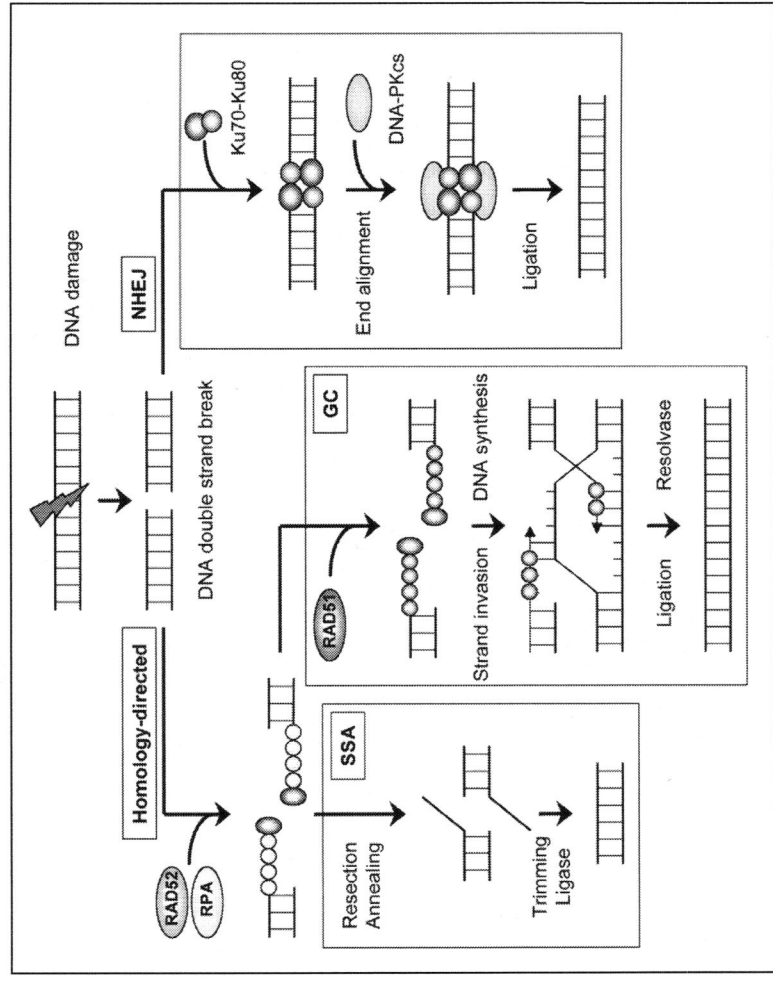

Role of *BRCA1* in DNA Repair

Involvement of BRCA1 in DNA Damage Signaling

Prior to the repair of DNA damage taking place, a cell must first recognise that DNA damage has occurred. In mammalian cells, several kinases, such as ATM, ATR, DNA-PK, Chk1 and Chk2, are activated in response to DNA damage. ATM (Ataxia Telangiectasia mutated) and ATR (Ataxia Telangectasia related) are central to the DNA damage response, relaying the signal onwards by phosphorylating downstream proteins. Following DNA damage, BRCA1 is rapidly phosphorylated by three distinct kinases, ATM, ATR and Chk2, which are activated by different stimuli and target distinct clusters of serine residues. Both ATM and Chk2 phosphorylate BRCA1 in response to ionizing radiation and these have been shown to be important for the BRCA1-mediated response to DNA damage.[45,46] ATR is also known to phosphorylate BRCA1, however this phosphorylation occurs in response to hydroxyurea-induced replication arrest or UV damage.[47] These phosphorylation events are used to modulate BRCA1 function to an extent that is not yet fully clear, but which probably results in the regulation of multiple downstream effectors.

Role of BRCA1 in the Repair of Double Strand Breaks

One of the earliest indications that BRCA1 was involved in DNA repair was the observation that BRCA1 associates and colocalizes with RAD51 in nuclear foci in mitotic and meiotic cells.[48] In meiotic cells, these foci were found to be associated with unsynapsed axial elements of developing synaptonemal complexes, implicating BRCA1 in genetic recombination events during meiosis. In mitotic cells, BRCA1/RAD51 foci were observed in S and G2 phases of the cell cycle. DNA damage induced by hydroxyurea or UV irradiation, caused the translocation of BRCA1/RAD51 foci to PCNA-containing replicating structures, indicating a role for BRCA1 in the repair of damaged replication forks.[48] BRCA2 and the BRCA1 binding protein, BARD1, were also shown to colocalize with BRCA1 and RAD51 in these foci, both before and after DNA damage.[48,49] This indicated that both BRCA1 and BRCA2 participated in the cellular response to DNA damage. However, it has been estimated that less than 5% of BRCA1 and BRCA2 form a complex with each other and a direct interaction between these two proteins has not been confirmed, although they can be coimmunoprecipitated.[49] Indeed, both BRCA1 and BRCA2 were recently identified as being part of a new protein complex called BRCC, which will be discussed further below.[50] A direct interaction between BRCA1 and RAD51 seems unlikely, whereas the direct interaction between BRCA2 and RAD51 is well established and will be discussed in greater detail below.

Further evidence for a role for BRCA1 in DNA repair came from the observation that BRCA1 deficient human tumour cells and mouse embryonic fibroblasts were highly sensitive to ionizing radiation[36,51] and displayed chromosomal instability, with both numerical and structural chromosome aberrations.[36,52-54] This indicated that BRCA1 defective cells were deficient in the repair of DSBs.

BRCA1 has been shown to be involved in HR,[37,55] but its involvement in NHEJ is still controversial. By using an artificial repair substrate integrated at a single site in the mouse genome it was shown that ES cells deficient in Brca1 had a 5-6 fold reduction in the repair of DSBs by HR but no defect in NHEJ.[55] However, using pulse-field gel electrophoresis, which is predominantly thought to reflect the NHEJ type of DNA DSB repair, cells deficient in BRCA1 have been reported to show defects in the rejoining of fragmented DNA after ionizing radiation.[51] However, other studies have followed with conflicting results.[56,57] Nevertheless, the involvement of BRCA1 in the repair of DSBs by HR is consistent with its association and colocalization with RAD51 in nuclear foci.

Other Repair Associated Functions of BRCA1

BRCA1 has been associated with a variety of proteins that are involved in the response or repair of DNA damage. The protein has been shown to colocalize with the RAD50-MRE11-NBS1 complex, which functions both in HR and NHEJ.[58] These proteins translocate to nuclear foci, independent of RAD51 foci, in response to DSBs induced by IR. However, BRCA1 does not seem to be required for the formation of these foci and therefore their significance is not fully clear.[59] It has been suggested that BRCA1 regulates the activity of the RAD50-MRE11-NBS1 complex by inhibiting the exonuclease activity of MRE11. This inhibition is the result of DNA binding by the BRCA1 protein, which exhibits a preference for branched DNA structures.[60] This has been suggested to favor DSB repair by HR at the expense of the more error-prone NHEJ pathway.

BRCA1 has been shown to be part of a large protein complex, termed BASC (BRCA1-associated genome surveillance complex).[61] This complex includes tumour suppressors, DNA damage sensors and signal transducers, including the RAD50-MRE11-NBS1 complex, the mismatch repair proteins MSH2, MSH6 and MLH1, the Blooms syndrome helicase BLM, the ATM kinase, DNA replication factor C (RFC) and PCNA. Most of these proteins can bind abnormal DNA structures and could thus act as sensors of DNA damage. Many also function directly in DNA replication or repair of replication associated DNA damage. This suggests a role for BRCA1 in coordinating various functions of DNA replication that are important for maintaining genomic integrity in the cell.

Whether the BRCA proteins have a role in NER is not clear. NER is considered to be involved in the repair of DNA cross-links along with HR[62] and thus the apparent sensitivity of BRCA deficient cells to DNA cross-linking agents could also be due to a defect in the NER pathway. BRCA1 and BRCA2 have both been implicated in transcription-coupled repair (TCR), a subpathway of NER.[63] However, the role of BRCA1 in this pathway is subject to reevalution following the retraction of a paper implicating BRCA1 in the repair of lesions due to oxidative damage.[64] BRCA1 has also been associated with another NER subpathway, global genomic repair (GGR), and was reported to enhance the GGR pathway by inducing the expression of the NER genes *XPC*, *DDB2* and *GADD45*.[65,66]

Recently yet another BRCA1 containing protein complex was identified, termed BRCC (BRCA1-BRCA2-containing complex)[50] that contains BRCA1, BRCA2, BARD1, RAD51 and other novel and unidentified components. This complex displays an E3 ubiquitin ligase activity, which has been implicated in the regulation of factors involved in DNA repair. In support of this, an increased association of RAD51 and p53 with the BRCC complex was observed following DNA damage and the C-terminal domain of p53 could be readily ubiquitinated by the BRCC complex.

Role of *BRCA2* in DNA Repair

The Role of BRCA2 in Homologous Recombination

BRCA1 and BRCA2 have been implicated in similar DNA repair processes through their colocalization along with RAD51 in nuclear foci. However, BRCA2 has not been detected in all BRCA1-associated protein complexes. The main role of BRCA2 appears to be to regulate the function of RAD51 in HR following DNA damage. BRCA2 has been shown to bind to RAD51 through six of the eight conserved BRC repeats encoded by exon 11, and through an unrelated site at the C-terminus.[23,34] These BRC repeats are well conserved through evolution, which indicates that they are functionally important. This has been confirmed in mouse model studies where retention of some or all of the BRC repeats are necessary for viability.[31]

It is now clear that BRCA2 is important for RAD51-dependant DNA repair of DSBs by HR. An initial indication of this, besides RAD51 binding, was that BRCA2 deficient cells were sensitive to ionizing radiation, which primarily causes DSBs.[34] Tumors, both from mice and humans, defective in BRCA2 also showed a high degree of chromosomal instability, including breaks and radial chromosomes,[33,67] which progressively accumulated during passage in culture.[68] These phenotypes indicated a defect in the recombinational repair of damaged DNA. BRCA2 deficient cells also fail to accumulate RAD51 in nuclear foci, which are thought to be the sites of DNA repair, following IR-induced DNA damage.[69] By using a chromosomally integrated repair substrate, BRCA2 deficient cells have been shown to exhibit a defect in the repair of DSBs by HR.[35,70] More specifically, BRCA2 has been shown to be important for the accurate repair of DSBs by gene conversion, while the NHEJ pathway seems to be unaffected. Interestingly, these cells also show an increase in the use of the more error-prone SSA pathway, which could explain the apparent chromosomal instability observed in BRCA2 deficient cells.[35]

It has now been established that the interaction between BRCA2 and RAD51 is necessary for the repair of DSBs by gene conversion, but how does BRCA2 affect RAD51 function? It has been suggested that BRCA2 is necessary for the transport of RAD51 into the nucleus to sites of DNA damage, where, in response to an appropriate signal, RAD51 would be released.[71] In support of this, truncated BRCA2, lacking its nuclear localization signals at the C-terminus, has been shown to be cytoplasmic along with the majority of the RAD51 protein in these cells.[71,72] BRCA2 could thus facilitate the transport of RAD51 into the nucleus following DNA damage, which would explain the impaired ability of BRCA2 deficient cells to form DNA damage-induced RAD51 foci. Overexpression of a single BRC repeat has also been shown to prevent the formation of RAD51 nucleoprotein filament formation, which is critical for HR to take place, by preventing RAD51 from binding to DNA.[71] A structural study has shown that the BRC repeat seems to mimic the structure of the interacting domain between RAD51 monomers, thereby impairing the ability of one RAD51 monomer to interact with another to form nucleoprotein filaments.[73] BRCA2 could thus prevent RAD51 from interacting with inappropriate DNA substrates and initiating a DNA damage response. These results indicate that BRCA2 could hold RAD51 in state of readiness in the cytoplasm until DNA damage occurs at which time BRCA2 delivers RAD51 into the nucleus to sites of DNA damage.

The C-terminus of BRCA2 has also been found to have a role in HR. This region shows high sequence conservation, even more so than the BRC repeats. It binds to a 70 amino acid protein called DSS1, whose amino acid sequence is 100% conserved between human and mouse.[22] The *DSS1* gene (deleted in split hand split foot syndrome) is one of three genes mapping to a 1.5Mb region, which is deleted in a rare developmental disorder of the limbs.[74] The structure of the mouse Brca2 C-terminal region has been solved bound to DSS1, and has been shown to have five distinct domains.[75] The first domain consists mostly of α-helices, followed by three oligonucleotide/oligosaccharide-binding (OB) folds (Fig. 1). Interestingly, the structure of the OB folds are known to be present in ssDNA binding proteins, such as Replication Protein A (RPA), and consistent with this the region was shown to be capable of binding ssDNA. The fifth domain is a tower-like structure extending from OB2, which has been inferred to bind dsDNA. DSS1 binds BRCA2 in an extended confirmation interacting with the helical domain, OB1 and OB2. It has been speculated that BRCA2 might be able to interact with ssDNA in the absence of DSS1, but other than this, little is known about the role of DSS1 in BRCA2-associated function. From these results, it has been suggested that BRCA2 recruits RAD51 to sites of DSBs by recognizing ssDNA/dsDNA structures present at processed break sites. This region was also shown to stimulate the DNA strand-exchange reaction promoted by RAD51 by an unknown mechanism, which possibly involves facilitating the loading of RAD51 onto ssDNA or RAD51 filament organization.[75]

Other Repair Associated Functions of BRCA2

Another conserved region identified in the *BRCA2* gene spans the region encoded by exons 2 and 3 at the N-terminus of the protein.[76] RPA, a ssDNA binding protein involved in DSB repair, has been reported to bind to the region encoded by exon 3 in the *BRCA2* gene, but the significance of this interaction is unknown.[77] This region was previously suggested to be involved in transcriptional activation[78] and recently, it has been reported to bind to a newly identified protein called EMSY.[79] This protein was reported to act as a repressor of the transactivation potential of BRCA2, although the direct consequences of this binding on BRCA2 function was not discussed.

BRCA2 has recently been reported to interact with the Polo-like kinase, Plk1.[80,81] Plk1-dependent phosphorylation of BRCA2 was observed during the G2/M phases of the cell cycle but was inhibited by DNA damage. The significance of this finding for BRCA2 function is not clear but it has been suggested that the phosphorylation might alter the spectrum of proteins bound to BRCA2.

BRCA1 and BRCA2 and Links to Other DNA Repair Syndromes

Mutations in the *BRCA1* and *BRCA2* genes predispose carriers to cancer through chromosome instability. Other genomic instability syndromes are well known, such as Ataxia Telangiectasia (AT), Bloom Syndrome (BS), Nijmegen Breakage Syndrome (NBS) and Fanconi anaemia (FA), which are all caused by a defective gene involved in the DNA damage response.[82] AT is a rare disorder characterized by progressive neuronal degeneration, immunodeficiency, cancer susceptibility, radiation sensitivity and telangiectasia of the face. It is caused by biallelic mutations in the *ATM* gene. BS is an autosomal recessive disorder caused by mutations in the *BLM* gene, which is a DNA helicase. Multiple systems are affected, characterized primarily by facial telangiectasia, short stature and a predisposition to cancers of all types. Cells from BS sufferers display extreme genomic instability. NBS is a rare autosomal recessive disorder, caused by mutations in the *NBS1* gene. Carriers suffer congenital abnormalities, immunodeficiency, chromosomal instability, X-ray hypersensitivity and a predisposition to malignancy. Evidence is gradually emerging of connections between many of these syndromes. As previously discussed, the ATM, BLM and NBS1 proteins have all been shown to associate with BRCA1. BRCA1 is phosphorylated by ATM in response to DNA damage and BRCA1 exists within the BASC complex in the nucleus together with ATM, BLM and NBS1. NBS1 is also part of the RAD50-MRE11-NBS1 complex, which has been shown to colocalize with BRCA1.

BRCA1 and *BRCA2* have also been linked to Fanconi anaemia (FA), a rare, autosomal recessive disease in which patients display wide clinical and genetic heterogeneity. Clinically the disease is characterized by progressive bone marrow failure, usually in the first decade of life, and increased susceptibility to malignancy, particularly acute myeloid leukaemia but also many other cancers such as squamous cell carcinoma of the head and neck.[83,84] On a cellular level, cells from patients with FA exhibit spontaneous chromosomal breakage and are hypersensitive to DNA crosslinking agents such as mitomycin C (MMC) and diepoxybutane (DEB). Following treatment with such crosslinking agents, FA cells display increased chromosome breakage and the formation of radial chromosomes, which is similar to the phenotype displayed by BRCA1 and BRCA2 deficient cells.

Eleven FA complementation groups have now been identified, *FANCA, B, C, D1, D2, E, F, G, L, I and J*,[85-87] all of which have been connected to distinct disease genes, except B and, the recently identified, I and J. Five of the FA proteins—A, C, E, F and G seem to interact in a multi-subunit nuclear complex, which assembles in response to DNA damage.[88] Subsequent to the formation of this complex, FANCD2 is monoubiquitinated and targeted to nuclear foci where it colocalizes with BRCA1 (Fig. 3). Monoubiquitination of FANCD2 requires an intact

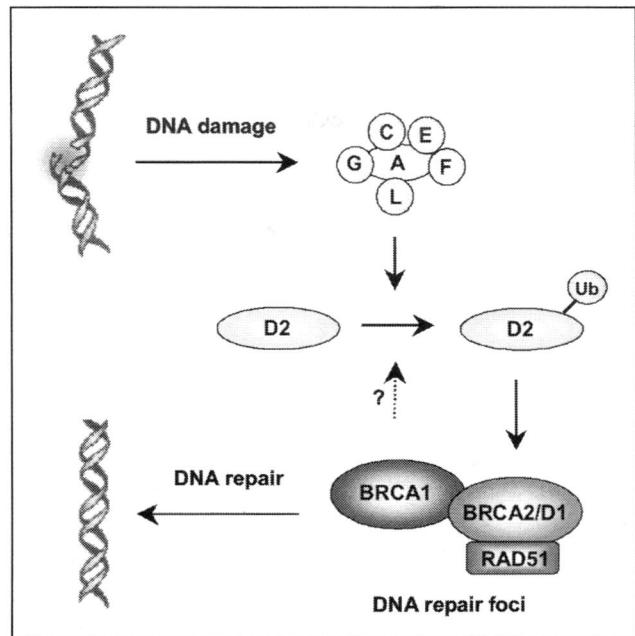

Figure 3. The Fanconi anemia/BRCA pathway. The Fanconi anemia proteins, A, C, E, F, G and L are thought to form a complex in the nucleus. Following DNA damage, FANCL monoubiquinates FANCD2, which in turn translocates to DNA repair foci where it co-localizes BRCA1. These foci are also thought to include BRCA2/FANCD1 and RAD51, which are important for the repair of DSBs by HR. FANCD2 has also been suggested as a substrate for the ubiquitin ligase activity of BRCA1.

nuclear complex but none of the 5 FA proteins identified within the complex contain a recognisable ubiquitin ligase motif. It was originally suggested that FANCD2 might be monoubiquitinated by BRCA1, since the BRCA1/BARD1 complex is known to have ubiquitin E3 ligase activity.[89] However, recent evidence has suggested that in vivo, monoubiquitinated FANCD2 exists at normal levels in the absence of BRCA1, but that its targeting to nuclear foci is under the control of BRCA1.[90] Recently a FA gene named *FANCL* (also known as *PHF9*) was identified and shown to be a component of the core nuclear complex.[87] It possesses ubiquitin ligase activity in vitro and appears to be required for FANCD2 monoubiquitination in vivo. BRCA1 has also been shown to interact directly with FANCA independent of DNA damage,[91] but the significance of this interaction is unknown.

Further links between FA and breast cancer became evident in 2002 when Howlett et al showed that the *FANCD1* gene is in fact *BRCA2*.[92] Biallelic mutations were found in the *BRCA2* gene in cells from patients assigned to complementation groups *D1* and *B* of FA. Expression of full length *BRCA2* cDNA was found to functionally complement the crosslinking hypersensitivity of *FANCD1* cells, suggesting that *BRCA2* and *FANCD1* are the same gene. Despite biallelic *BRCA2* mutations also being found in patients from the *FANCB* complementation group, the *BRCA2* cDNA did not complement the cellular hypersensitivity phenotype and therefore the evidence for *BRCA2* being *FANCB* is less convincing.

FANCG was recently reported to interact with FANCD1/BRCA2 and colocalize in nuclear foci with both BRCA2 and RAD51 following DNA damage.[93] FANCG is part of the core nuclear complex but evidence so far suggests that FANCD1/BRCA2 functions downstream of this complex, since monoubiquitination of FANCD2 and targeting to nuclear foci is normal in cells from patients from complementation group D1. It would thus seem likely that this interaction is independent to the participation of FANCG within the core nuclear complex. Recently, a reduction in HR in *FANCG* deficient DT40 chicken cells was demonstrated,[94] which is similar to that seen in *BRCA1* and *BRCA2* deficient cells.[35,37] The same cells had normal levels of RAD51 focus formation after DNA damage with IR and MMC. Indeed, there is currently conflicting evidence concerning RAD51 foci formation in FA cells. One report suggests DNA damage-induced RAD51 foci are attenuated in cells from all Fanconi complementation groups,[95] while another suggests RAD51 foci are only reduced in cells from complementation group *FANCD1/BRCA2*.[96]

It is widely accepted that the components of the FA pathway function within the cellular DNA repair mechanism. Exactly how this occurs and at what stage within the DNA damage response is unclear. However, the convergence of these two previously unrelated pathways—FA and BRCA1/BRCA2—is proving to be an interesting development in the field of DNA repair.

Clinical and Therapeutic Implications

With the advent of increased knowledge and understanding about the function of the BRCA1 and BRCA2 proteins and their involvement in tumorigenesis, it is hoped that this will lead to the development of breast cancer treatment tailored more precisely to target the specificities of tumors in *BRCA1* and *BRCA2* mutation carriers. Since the wild-type allele is inactivated in the tumors of people with inherited germ-line mutations in *BRCA1* and *BRCA2*, it would therefore follow that treatments directed specifically to kill cells with biallelic mutations in either gene should serve to be lethal to the tumour cells only, while sparing the surrounding, normal tissue. Since cells with no functional BRCA1 or BRCA2 are known to be hypersensitive to DNA damaging agents that cause DSBs, such as the DNA crosslinking agent cisplatin, these agents could be used to drive these cells into apoptosis, while normal cells should be able to repair the breaks. Another way to achieve this would be to render the remaining normal repair pathways in the BRCA deficient cells nonfunctional, further compromising their ability to repair DNA damage and therefore their viability. BRCA1 and BRCA2 defective cells are known to be deficient in the repair of DSBs by HR, while other repair pathways seem to be largely unaffected. These pathways could therefore be used as therapeutic targets in BRCA tumors.

References

1. Collins N, McManus R, Wooster R et al. Consistent loss of the wild type allele in breast cancers from a family linked to the BRCA2 gene on chromosome 13q12-13. Oncogene 1995; 10:1673-1675.
2. Cornelis RS, Neuhausen SL, Johansson O et al. High allele loss rates at 17q12-q21 in breast and ovarian tumors from BRCAl-linked families. The Breast Cancer Linkage Consortium. Genes Chromosomes Cancer 1995; 13:203-210.
3. Ford D, Easton DF, Stratton M et al. Genetic heterogeneity and penetrance analysis of the BRCA1 and BRCA2 genes in breast cancer families. The Breast Cancer Linkage Consortium. Am J Hum Genet 1998; 62:676-689.
4. Ford D, Easton DF, Bishop DT et al. Risks of cancer in BRCA1-mutation carriers. Breast Cancer Linkage Consortium. Lancet 1994; 343:692-695.
5. Cancer risks in BRCA2 mutation carriers.The Breast Cancer Linkage Consortium. J Natl Cancer Inst 1999; 91:1310-1316.
6. Narod SA. Modifiers of risk of hereditary breast and ovarian cancer. Nat Rev Cancer 2002; 2:113-123.
7. Esteller M, Silva JM, Dominguez G et al. Promoter hypermethylation and BRCA1 inactivation in sporadic breast and ovarian tumors. J Natl Cancer Inst 2000; 92:564-569.

8. Welcsh PL,King MC. BRCA1 and BRCA2 and the genetics of breast and ovarian cancer. Hum Mol Genet 2001; 10:705-713.

9. Hilton JL, Geisler JP, Rathe JA et al. Inactivation of BRCA1 and BRCA2 in ovarian cancer. J Natl Cancer Inst 2002; 94:1396-1406.

10. Collins N, Wooster R,Stratton MR. Absence of methylation of CpG dinucleotides within the promoter of the breast cancer susceptibility gene BRCA2 in normal tissues and in breast and ovarian cancers. Br J Cancer 1997; 76:1150-1156.

11. Lakhani SR, Gusterson BA, Jacquemier J et al. The pathology of familial breast cancer: Histological features of cancers in families not attributable to mutations in BRCA1 or BRCA2. Clin Cancer Res 2000; 6:782-789.

12. Miki Y, Swensen J, Shattuck-Eidens D et al. A strong candidate for the breast and ovarian cancer susceptibility gene BRCA1. Science 1994; 266:66-71.

13. Hall JM, Lee MK, Newman B et al. Linkage of early-onset familial breast cancer to chromosome 17q21. Science 1990; 250:1684-1689.

14. Hashizume R, Fukuda M, Maeda I et al. The RING heterodimer BRCA1-BARD1 is a ubiquitin ligase inactivated by a breast cancer-derived mutation. J Biol Chem 2001; 276:14537-14540.

15. Chen A, Kleiman FE, Manley JL et al. Autoubiquitination of the BRCA1*BARD1 RING ubiquitin ligase. J Biol Chem 2002; 277:22085-22092.

16. Manke IA, Lowery DM, Nguyen A et al. BRCT repeats as phosphopeptide-binding modules involved in protein targeting. Science 2003; 302:636-639.

17. Yu X, Chini CC, He M et al. The BRCT domain is a phospho-protein binding domain. Science 2003; 302:639-642.

18. Kerr P, Ashworth A. New complexities for BRCA1 and BRCA2. Curr Biol 2001; 11:R668-676.

19. Wooster R, Neuhausen SL, Mangion J et al. Localization of a breast cancer susceptibility gene, BRCA2, to chromosome 13q12-13. Science 1994; 265:2088-2090.

20. Wooster R, Bignell G, Lancaster J et al. Identification of the breast cancer susceptibility gene BRCA2. Nature 1995; 378:789-792.

21. Bork P, Blomberg N, Nilges M. Internal repeats in the BRCA2 protein sequence. Nat Genet 1996; 13:22-23.

22. Marston NJ, Richards WJ, Hughes D et al. Interaction between the product of the breast cancer susceptibility gene BRCA2 and DSS1, a protein functionally conserved from yeast to mammals. Mol Cell Biol 1999; 19:4633-4642.

23. Wong AK, Pero R, Ormonde PA et al. RAD51 interacts with the evolutionarily conserved BRC motifs in the human breast cancer susceptibility gene brca2. J Biol Chem 1997; 272:31941-31944.

24. Gudas JM, Li T, Nguyen H et al. Cell cycle regulation of BRCA1 messenger RNA in human breast epithelial cells. Cell Growth Differ 1996; 7:717-723.

25. Bertwistle D, Swift S, Marston NJ et al. Nuclear location and cell cycle regulation of the BRCA2 protein. Cancer Res 1997; 57:5485-5488.

26. Blackshear PE, Goldsworthy SM, Foley JF et al. Brca1 and Brca2 expression patterns in mitotic and meiotic cells of mice. Oncogene 1998; 16:61-68.

27. Rajan JV, Marquis ST, Gardner HP et al. Developmental expression of Brca2 colocalizes with Brca1 and is associated with proliferation and differentiation in multiple tissues. Dev Biol 1997; 184:385-401.

28. Rajan JV, Wang M, Marquis ST et al. Brca2 is coordinately regulated with Brca1 during proliferation and differentiation in mammary epithelial cells. Proc Natl Acad Sci USA 1996; 93:13078-13083.

29. Moynahan ME. The cancer connection: BRCA1 and BRCA2 tumor suppression in mice and humans. Oncogene 2002; 21:8994-9007.

30. Hakem R, de la Pompa JL, Elia A et al. Partial rescue of Brca1 (5-6) early embryonic lethality by p53 or p21 null mutation. Nat Genet 1997; 16:298-302.

31. Connor F, Bertwistle D, Mee PJ et al. Tumorigenesis and a DNA repair defect in mice with a truncating Brca2 mutation. Nat Genet 1997; 17:423-430.

32. Ludwig T, Chapman DL, Papaioannou VE et al. Targeted mutations of breast cancer susceptibility gene homologs in mice: Lethal phenotypes of Brca1, Brca2, Brca1/Brca2, Brca1/p53, and Brca2/p53 nullizygous embryos. Genes Dev 1997; 11:1226-1241.

33. Yu VP, Koehler M, Steinlein C et al. Gross chromosomal rearrangements and genetic exchange between nonhomologous chromosomes following BRCA2 inactivation. Genes Dev 2000; 14:1400-1406.
34. Sharan SK, Morimatsu M, Albrecht U et al. Embryonic lethality and radiation hypersensitivity mediated by Rad51 in mice lacking Brca2. Nature 1997; 386:804-810.
35. Tutt A, Bertwistle D, Valentine J et al. Mutation in Brca2 stimulates error-prone homology-directed repair of DNA double-strand breaks occurring between repeated sequences. Embo J 2001; 20:4704-4716.
36. Shen SX, Weaver Z, Xu X et al. targeted disruption of the murine Brca1 gene causes gamma-irradiation hypersensitivity and genetic instability. Oncogene 1998; 17:3115-3124.
37. Moynahan ME, Cui TY, Jasin M. Homology-directed dna repair, mitomycin-c resistance, and chromosome stability is restored with correction of a Brca1 mutation. Cancer Res 2001; 61:4842-4850.
38. Hoeijmakers JH. Genome maintenance mechanisms for preventing cancer. Nature 2001; 411:366-374.
39. Khanna KK, Jackson SP. DNA double-strand breaks: Signaling, repair and the cancer connection. Nat Genet 2001; 27:247-254.
40. van Gent DC, Hoeijmakers JH, Kanaar R. Chromosomal stability and the DNA double-stranded break connection. Nat Rev Genet 2001; 2:196-206.
41. Valerie K, Povirk LF. Regulation and mechanisms of mammalian double-strand break repair. Oncogene 2003; 22:5792-5812.
42. Liang F, Han M, Romanienko PJ et al. Homology-directed repair is a major double-strand break repair pathway in mammalian cells. Proc Natl Acad Sci USA 1998; 95:5172-5177.
43. Baumann P, West SC. Role of the human RAD51 protein in homologous recombination and double- stranded-break repair. Trends Biochem Sci 1998; 23:247-251.
44. Richardson C, Jasin M. Frequent chromosomal translocations induced by DNA double-strand breaks. Nature 2000; 405:697-700.
45. Cortez D, Wang Y, Qin J et al Requirement of ATM-dependent phosphorylation of brca1 in the DNA damage response to double-strand breaks. Science 1999; 286:1162-1166.
46. Lee JS, Collins KM, Brown AL et al. hCds1-mediated phosphorylation of BRCA1 regulates the DNA damage response. Nature 2000; 404:201-204.
47. Tibbetts RS, Cortez D, Brumbaugh KM et al. Functional interactions between BRCA1 and the checkpoint kinase ATR during genotoxic stress. Genes Dev 2000; 14:2989-3002.
48. Scully R, Chen J, Plug A et al. Association of BRCA1 with Rad51 in mitotic and meiotic cells. Cell 1997; 88:265-275.
49. Chen J, Silver DP, Walpita D et al. Stable interaction between the products of the BRCA1 and BRCA2 tumor suppressor genes in mitotic and meiotic cells. Mol Cell 1998; 2:317-328.
50. Dong Y, Hakimi MA, Chen X et al. Regulation of BRCC, a holoenzyme complex containing BRCA1 and BRCA2, by a signalosome-like subunit and its role in DNA repair. Mol Cell 2003; 12:1087-1099.
51. Scully R, Ganesan S, Vlasakova K et al. Genetic analysis of BRCA1 function in a defined tumor cell line. Mol Cell 1999; 4:1093-1099.
52. Xu X, Wagner KU, Larson D et al. Conditional mutation of Brca1 in mammary epithelial cells results in blunted ductal morphogenesis and tumour formation. Nat Genet 1999; 22:37-43.
53. Weaver Z, Montagna C, Xu X et al. Mammary tumors in mice conditionally mutant for Brca1 exhibit gross genomic instability and centrosome amplification yet display a recurring distribution of genomic imbalances that is similar to human breast cancer. Oncogene 2002; 21:5097-5107.
54. Tirkkonen M, Johannsson O, Agnarsson BA et al. Distinct somatic genetic changes associated with tumor progression in carriers of BRCA1 and BRCA2 germ-line mutations. Cancer Res 1997; 57:1222-1227.
55. Moynahan ME, Chiu JW, Koller BH et al. Brca1 controls homology-directed DNA repair. Mol Cell 1999; 4:511-518.
56. Wang H, Zeng ZC, Bui TA et al. Nonhomologous end-joining of ionizing radiation-induced DNA double- stranded breaks in human tumor cells deficient in BRCA1 or BRCA2. Cancer Res 2001; 61:270-277.

57. Zhong Q, Boyer TG, Chen PL et al. Deficient nonhomologous end-joining activity in cell-free extracts from Brca1-null fibroblasts. Cancer Res 2002; 62:3966-3970.

58. Zhong Q, Chen CF, Li S et al. Association of BRCA1 with the hRad50-hMre11-p95 complex and the DNA damage response. Science 1999; 285:747-750.

59. Wu X, Petrini JH, Heine WF et al. Independence of R/M/N focus formation and the presence of intact BRCA1. Science 2000; 289:11.

60. Paull TT, Cortez D, Bowers B et al. Direct DNA binding by Brca1. Proc Natl Acad Sci USA 2001; 98:6086-6091.

61. Wang Y, Cortez D, Yazdi P et al. BASC, a super complex of BRCA1-associated proteins involved in the recognition and repair of aberrant DNA structures. Genes Dev 2000; 14:927-939.

62. De Silva IU, McHugh PJ, Clingen PH et al. Defining the roles of nucleotide excision repair and recombination in the repair of DNA interstrand cross-links in mammalian cells. Mol Cell Biol 2000; 20:7980-7990.

63. Le Page F, Randrianarison V, Marot D et al. BRCA1 and BRCA2 are necessary for the transcription-coupled repair of the oxidative 8-oxoguanine lesion in human cells. Cancer Res 2000; 60:5548-5552.

64. Gowen LC, Avrutskaya AV, Latour AM et al. BRCA1 required for transcription-coupled repair of oxidative DNA damage. Science 1998; 281:1009-1012.

65. Hartman AR, Ford JM. BRCA1 induces DNA damage recognition factors and enhances nucleotide excision repair. Nat Genet 2002; 32:180-184.

66. Harkin DP, Bean JM, Miklos D et al. Induction of GADD45 and JNK/SAPK-dependent apoptosis following inducible expression of BRCA1. Cell 1999; 97:575-586.

67. Gretarsdottir S, Thorlacius S, Valgardsdottir R et al. BRCA2 and p53 mutations in primary breast cancer in relation to genetic instability. Cancer Res 1998; 58:859-862.

68. Patel KJ, Vu VP, Lee H et al. Involvement of Brca2 in DNA repair. Mol Cell 1998; 1:347-357.

69. Yuan SS, Lee SY, Chen G et al. BRCA2 is required for ionizing radiation-induced assembly of Rad51 complex in vivo. Cancer Res 1999; 59:3547-3551.

70. Moynahan ME, Pierce AJ, Jasin M. BRCA2 is required for homology-directed repair of chromosomal breaks. Mol Cell 2001; 7:263-272.

71. Davies AA, Masson JY, McIlwraith MJ et al. Role of BRCA2 in control of the RAD51 recombination and DNA repair protein. Mol Cell 2001; 7:273-282.

72. Spain BH, Larson CJ, Shihabuddin LS et al. Truncated BRCA2 is cytoplasmic: Implications for cancer-linked mutations. Proc Natl Acad Sci USA 1999; 96:13920-13925.

73. Pellegrini L, Yu DS, Lo T et al. Insights into DNA recombination from the structure of a RAD51-BRCA2 complex. Nature 2002; 420:287-293.

74. Crackower MA, Scherer SW, Rommens JM et al. Characterization of the split hand/split foot malformation locus SHFM1 at 7q21.3-q22.1 and analysis of a candidate gene for its expression during limb development. Hum Mol Genet 1996; 5:571-579.

75. Yang H, Jeffrey PD, Miller J et al. BRCA2 function in DNA binding and recombination from a BRCA2-DSS1-ssDNA structure. Science 2002; 297:1837-1848.

76. Warren M, Smith A, Partridge N et al. Structural analysis of the chicken BRCA2 gene facilitates identification of functional domains and disease causing mutations. Hum Mol Genet 2002; 11:841-851.

77. Wong JM, Ionescu D, Ingles CJ. Interaction between BRCA2 and replication protein a is compromised by a cancer-predisposing mutation in BRCA2. Oncogene 2003; 22:28-33.

78. Milner J, Ponder B, Hughes-Davies L et al. Transcriptional activation functions in BRCA2. Nature 1997; 386:772-773.

79. Hughes-Davies L, Huntsman D, Ruas M et al. EMSY links the BRCA2 pathway to sporadic breast and ovarian cancer. Cell 2003; 115:523-535.

80. Lee M, Daniels MJ, Venkitaraman AR. Phosphorylation of BRCA2 by the Polo-like kinase Plk1 is regulated by DNA damage and mitotic progression. Oncogene 2003.

81. Lin HR, Ting NS, Qin J. M phase-specific phosphorylation of BRCA2 by Polo-like kinase 1 correlates with the dissociation of the BRCA2-P/CAF complex. J Biol Chem 2003; 278:35979-35987.

82. Thompson LH, Schild D. Recombinational DNA repair and human disease. Mutat Res 2002; 509:49-78.

83. Joenje H, Patel KJ. The emerging genetic and molecular basis of Fanconi anaemia. Nat Rev Genet 2001; 2:446-457.

84. D'Andrea AD, Grompe M. The Fanconi anaemia/BRCA pathway. Nat Rev Cancer 2003; 3:23-34.

85. Joenje H, Oostra AB, Wijker M et al. Evidence for at least eight Fanconi anemia genes. Am J Hum Genet 1997; 61:940-944.

86. Levitus M, Rooimans MA, Steltenpool J et al. Heterogeneity in Fanconi anemia: Evidence for 2 new genetic subtypes. Blood 2004; 103:2498-2503.

87. Meetei AR, de Winter JP, Medhurst AL et al. A novel ubiquitin ligase is deficient in Fanconi anemia. Nat Genet 2003; 35:165-170.

88. Medhurst AL, Huber PA, Waisfisz Q et al. Direct interactions of the five known Fanconi anaemia proteins suggest a common functional pathway. Hum Mol Genet 2001; 10:423-429.

89. Garcia-Higuera I, Taniguchi T, Ganesan S et al. Interaction of the Fanconi anemia proteins and BRCA1 in a common pathway. Mol Cell 2001; 7:249-262.

90. Vandenberg CJ, Gergely F, Ong CY et al. BRCA1-independent ubiquitination of FANCD2. Mol Cell 2003; 12:247-254.

91. Folias A, Matkovic M, Bruun D et al. BRCA1 interacts directly with the Fanconi anemia protein FANCA. Hum Mol Genet 2002; 11:2591-2597.

92. Howlett NG, Taniguchi T, Olson S et al. Biallelic inactivation of BRCA2 in Fanconi anemia. Science 2002; 297:606-609.

93. Hussain S, Witt E, Huber PA et al. Direct interaction of the Fanconi anaemia protein FANCG with BRCA2/FANCD1. Hum Mol Genet 2003; 12:2503-2510.

94. Yamamoto K, Ishiai M, Matsushita N et al. Fanconi anemia FANCG protein in mitigating radiation- and enzyme-induced DNA double-strand breaks by homologous recombination in vertebrate cells. Mol Cell Biol 2003; 23:5421-5430.

95. Digweed M, Rothe S, Demuth I et al. Attenuation of the formation of DNA-repair foci containing RAD51 in Fanconi anaemia. Carcinogenesis 2002; 23:1121-1126.

96. Godthelp BC, Artwert F, Joenje H et al. Impaired DNA damage-induced nuclear Rad51 foci formation uniquely characterizes Fanconi anemia group D1. Oncogene 2002; 21:5002-5005.

CHAPTER 5

Radiosensitivity of Cells Derived from Down Syndrome Patients:
Is Defective DNA Repair Involved?

Adayapalam T. Natarajan

Abstract

Down's syndrome (DS) is an autosomal recessive human disorder caused by an extra copy of chromosome 21. DS patients are characterized by dwarfism and mental retardation accompanied by an increased incidence of cancer development in various tissues and organs. DS patients also show signs of premature aging phenotypes. Many of the phenotypic features of DS patients are presumably due to the excess of genetic material of chromosome 21. Cells derived from DS patients show abnormal response to ionizing radiation-induced DNA damage, and this review deals with some aspects of the radiosensitive phenotype in DS.

Introduction

Down syndrome (DS) is an autosomal recessive human hereditary disorder caused by an extra copy of chromosome 21. It is the most common postnatally viable chromosomal anomaly having an incidence of about 1 in 700 live births. DS is characterized by diverse developmental abnormalities facial dysmorphology, congenital defects of heart and gut, infertility and immunodeficiencies. In addition to mental retardation, DS patients have an increased cancer incidence in various organs and tissues and show signs of premature aging. Bone marrow dysfunction is one of the several disturbances seen in DS patients. Leukemia development is one of the major hematological abnormalities in DS patients, with 10 to 30 fold increases in occurrence. The underlying mechanisms for this increase are poorly understood.

Korenberg et al[1] have made a comprehensive review of the Down syndrome phenotype on the basis of chromosomal imbalance. The extra genetic material due to three copies of chromosome 21 may cause genetic imbalance resulting in an altered response to genetic and environmental factors. One such response is the increased sensitivity to ionizing radiation and this aspect will be reviewed in this paper.

Chromosome Aberrations in DS Cells

Frequencies of spontaneously occurring chromosomal aberrations in cells of DS origin remain the same in comparison to cells from normal healthy individuals. Choudina et al[2] and Dekaban et al[3] first reported an increased frequency of chromosomal aberrations induced by

DNA Repair and Human Disease, edited by Adayabalam S. Balajee. ©2006 Landes Bioscience and Springer Science+Business Media.

ionizing radiation in the lymphocytes of DS patients in 1966. This observation was confirmed by subsequent studies.[4,5] DS cells have also been shown to be sensitive to certain chemical clastogens of different types, such as bleomycin and alkylating agents [6,7,8] as well as exposure to viruses in vivo.[9,10] Most of the studies have been carried out by employing nonstimulated peripheral blood lymphocytes, irradiated at G0 stage. The yield of exchange type of aberrations, i.e, dicentrics is 1.5 to 2- fold higher in lymphocytes from DS patients compared to those from normal individuals. This increase is more predominant at higher doses of X-rays (<2 Gy). Recent studies using fluorescence in situ hybridization (FISH) technique employing chromosome specific DNA probes for 1, 4, 21 and 22 did not show any differences in the yield of X-ray induced translocations between the normal and DS lymphocytes.[11] The increased sensitivity of DS lymphocytes has also been observed following irradiation in G2 stage.[12,13]

Sister Chromatid Exchanges (SCEs) in DS Cells

Sister chromatid exchanges are formed during replication of chromosomes and occur spontaneously at very low frequencies, around one per genome per cell cycle in mammalian cells. Increased frequencies of spontaneous SCEs mainly arise due to incorporated 5-bromodeoxyuridine (BrdUrd) in the template strand of DNA.[14] There are conflicting reports on the frequencies of SCEs in DS lymphocytes. Higher, similar or lower frequencies in DS cells compared to controls have been reported [15] Since several factors, such as the extent of incorporation of BrdUrd, the length of cell cycle and culture conditions can potentially modulate the frequencies of observed SCEs, it is difficult to compare the results of different studies. However, an extensive study by Shubber et al.[15] indicates an increase in the frequencies of SCEs in DS infants and their parents in comparison to the normal infants and their parents. They also reported an enhanced response of DS lymphocytes to clastogenic agents, such as mitomycin C, Hycanthone and X-rays in comparison to lymphocytes from controls.

Mutations in Cultured DS Cells

Mutation frequencies (Mf) in the HPRT locus have been estimated in lymphocytes of infant and adult DS patients using the cloning method.[16] Unlike normal control population, in which the Mf increases with age, in DS patients such age dependency is not evident demonstrating that the frequency of spontaneous somatic mutations in children and adults with DS are atypical compared to normal controls and suggesting that the genetic mechanisms associated with background mutation in children and adult DS may be different.[16] Studies on the induction of mutations in HPRT locus in DS fibroblasts following X-irradiation have shown no difference between DS cells and normal cells either for spontaneous or induced mutation frequencies.[17]

Repair of DNA Damage in DS Cells

The increased sensitivity of DS cells to ionizing radiation has naturally led to studies to evaluate their ability to repair induced DNA damage. Ionizing radiation induces in cellular DNA, single strand breaks (SSB), double strand breaks (DSB), base damage (BD) and multiple lesions. DSB is considered to be the most important DNA lesion responsible for radiobiological effects including chromosomal aberrations.[18,19] There are only few studies, which have attempted to detect any defect in the repair of ionizing radiation induced DNA damage in DS cells. One of the difficulties to directly link the hyper- radiosensitivity observed as induced chromosomal aberrations in DS cells to any DNA repair defect is that most of the cytogenetic studies have been carried out with the peripheral blood lymphocytes, whereas the repair studies have been done in cultured fibroblasts. There is some evidence that the lymphocytes and fibroblasts of DS patients respond differently for induction of chromosomal aberrations by ionizing radiation.[20] One technical limitation for quantitative evaluation of DNA repair capacity of strand breaks in lymphocytes is that these cells cannot be prelabeled with radioactive precursors, which is often a prerequisite for most of the DNA repair studies. However, single cell gel electrophoresis assay, which does not require any prelabeling has not yet been utilized for DNA repair studies in DS lymphocytes

Repair of DNA Single Strand Breaks in DS Cells

Studying the DNA sedimentation profiles in alkaline sucrose gradients, Athanasiou et al.[21] determined that cultured lymphocytes from DS patients are less efficient in repairing X-ray induced single strand breaks. However, a later study using a more sensitive technique, namely alkaline unwinding in combination with S1 endonuclease, showed no difference in the repair of DNA single strand breaks in γ irradiated cultured lymphocytes from DS patients.[22] Steiner and Woods [23] evaluated the repair of γ rays induced single strand breaks (SSBs) and double strand breaks (DSBs) in nontransformed fibroblasts of DS patients and normal controls, using alkaline elution (SSB) and nondenaturing elution (DSB) and found no defects in DS cells for repair of both these types of lesions. Employing alkaline elution technique, the induction and repair of γ-rays induced SSB were studied in cultured DS lymphocytes and compared with lymphocytes from normal individuals by Chiriocolo et al.[24] These authors however found that DS lymphocytes repaired the strand breaks at a faster rate than the normal ones. Zinc supplementation to the DS children restored to a repair pattern similar to normal children.

Thus, there are conflicting reports about the repair defect in ionizing radiation induced DNA strand breaks in DS cells despite the use of similar techniques employed. The basis for these conflicting results is not clearly understood now. It possibly reflects the genetic heterogeneity of cells obtained from different DS patients. Due to these conflicting reports on the repair fidelity of strand breaks in DS cells, attempts to correlate between repair deficiency and chromosomal aberrations have proven difficult. Nevertheless it has been consistently observed that higher frequencies of aberrations are induced in DS lymphocytes irrespective of different cell cycle phases. One of the suggestions made to explain the increased sensitivity of DS cells is that the strand breaks are repaired faster in these cells and this allows an increased probability for mis-repair of breaks leading to chromosome aberrations.[25,26] It is obvious that further parallel studies using the same DS cells should be carried out measuring both DNA damage and induced chromosomal aberrations utilizing more sensitive techniques, such as single cell gel electrophoresis for repair and premature chromosome condensation for chromosome aberrations.

Unscheduled DNA Synthesis

When cells, not in DNA synthetic phase are irradiated with ultraviolet light or treated with chemical mutagens evoke unscheduled DNA synthesis during "long patch" repair that can be measured by incorporation of tritiated thymidine in the presence of hydroxyurea. Both fibroblasts and lymphocytes have been found to have a marked reduction in unscheduled DNA synthesis when compared to cells from normal individuals indicating a possible defect in nucleotide excision repair in DS cells.[27,28]

Radio-Resistant DNA Synthesis

Cultured cells from highly radiosensitive syndrome ataxia telangiectasia (AT) are abnormally resistant to the inhibition of DNA synthesis after exposure to ionizing radiation. Cells derived from DS patients, though sensitive to ionizing radiation as measured by cell killing and induced chromosome aberrations, however, were found to have normal inhibition of DNA synthesis, indicating radio-sensitivity and radio-resistant DNA synthesis are not always correlated.[29]

Repair of Mitochondrial DNA Damage

Oxidative DNA damage has been suspected to be one of the causal factors for aging. A number of age related neuromuscular degenerative diseases have been associated with mutations in mitochondrial DNA (mt DNA). Down syndrome is characterized by premature aging. When fibroblasts from DS patients were treated with menadione (a reactive oxygen generator) and the oxidative damage was measured at several time points after treatment, it was found that these cells have decreased ability to repair oxidative dame in mtDNA compared to age matched control cells.[30] Oxidative damage in nuclear DNA (nDNA) of post-mortom brain tissue has been evaluated by measuring the levels of 8-OhdG. There was no evidence for increased oxidative nDNA damage in brain cells of DS patients in comparison to brain cells from normal individuals.[31] Increased levels of expression of DNA repair genes, such as ERCC2, ERCC3, and XRCC1 have been found in different regions of brain from DS patients indicating a possible permanent oxidative DNA damage.[32] The significance of these findings has yet to be evaluated.

Mouse Models of DS

Mouse models have been used to determine the genotype-phenotype correlations in DS.[33] Mice with segmental trisomy for the distal end of the mouse chromosome 16, which genetically corresponds to most of the human chromosome 21, serve as a genetic model for DS.[34] However, one major limitation is the partial trisomy of mouse 16, which contains only a few comparable genes with humans. A recent study by Shinohara et al.[35] using a chimeric mice containing a human chromosome 21 displayed a high degree of correlation between the retention of chromosome 21 in the brain and its impairment assayed by learning or emotional behavior by open-field, contextual fear conditioning and forced swim tests. Chimeric mouse fetuses with a high retention of human chromosome 21 also showed hypoplastic thymus and cardiac defects. Sequencing of 99.7% of the long arm of human chromosome 21 revealed 127 known genes and 98 predicted genes. Knowledge of the genetic map of chromosome 21 has facilitated the research into the molecular mechanisms for DS features.

Recent studies have revealed that chromosome 21 contains 9 genes, which could possibly affect brain development, neuronal loss and Alzheimer's type neuropathology. It also contains 16 genes with known functions in energy and reaction oxygen species metabolism.[36] The question of how an extra copy of chromosome 21 can lead to diverse phenotypic traits of DS is currently under investigation. It is likely that over expression of proteins encoded by genes of chromosome 21 somehow interfere with normal cellular processes in different tissues and organs. An equally interesting possibility is that the spatial organization of interphase nuclear architecture is greatly affected by the presence of an extra copy of chromosome 21 which might result in the modulation of diverse DNA metabolic activities of replication, transcription, repair and recombination.

Conclusions

Cells derived from DS patients consistently exhibit increased radiosensitivity as measured by the induced frequencies of chromosomal aberrations. However, few studies carried out so far on the DNA repair capacity of DS cells have shown conflicting results. It is suggested that further studies on DS cells should be carried out in parallel using sensitive techniques such as single cell gel electrophoresis (for repair of strand breaks) and premature chromosome condensation (for chromosome aberrations) to establish a relationship between DNA repair and chromosomal aberrations induced by ionizing radiation. Future research will determine whether some of the phenotypic features associated with DS are due to deficiencies in DNA repair.

References

1. Korenberg JR, Chen K, Schipper R et al. Down syndrome phenotype: The consequences of chromosome imbalance. Proc Natl Acad Sci USA 1994; 91:4997-5001.
2. Chudina AP, Malyutina TS, Progosyane HE. Comparative radiosensitivity of chromosomes in the cultured peripheral blood lymphocytes of normal donors and patients with Down's syndrome. Genetika 1966; 4:51-63.
3. Dekaban ASR, Thron R, Streusing JK. Chromosomal aberrations in irradiated blood and blood cultures of normal subjects and selected patients with chromosomal abnormality. Radiat Res 1966; 27:50-63.
4. Kucherova M. Comparison of radiation effects in vitro up on chromosomes of human subjects. Acta Radiol 1967; 6:441-448.
5. Sasaki MS, Tonomura A. Chromosomal radiosensitivity in Down's syndrome. Jpn J Hum Genet 1969; 14:81-92.
6. O'Brien RL, Poon P, Kline E et al. Susceptibility of chromosomes from patients with Down's syndrome to 7,12-dimethylbenz(a)anthracene-induced aberrations in vitro. Int J Cancer 1971; 8:200-210.
7. Vijayalaxmi, Evans HJ. Bleomycin-induced chromosome aberrations in Down's syndrome lymphocytes. Mutation Res 1982; 105:107-113.
8. Kaina B, Waller H, Waller M et al. The action of N-methyl-N-nitrosourea on nonestablished human cell lines in vitro.I. Cell cycle inhibition and aberration induction in diploid and Down's fibroblasts. Mutation Res 1997; 43:387-400.
9. Higurashi M, Tamura T, Nakatake T. Cytogenetic observations in cultured lymphocytes from patients with Down's syndrome and measles, Pediat. Res 1973; 7:582-587.
10. Higurashi M, Tada A, Miyahara S et al. Chromosome damage in Down's syndrome by chicken pox infection. Pediat Res 1976; 10:189-192.
11. Grigorova M, Natarajan AT. Relative involvement of chromosome #21 in radiation induced exchange aberrations in lymphocytes of Down syndrome patients. Mutation Res 1998; 404:67-75.
12. Natarajan AT, G Obe, FN Dulout. The effect of caffeine post-treatment on X-ray induced chromosomal aberrations in human blood lymphocytes in vitro. Human.Gene0t 1980; 54:183-189.
13. Picheira J, Rodriguez M, Bravo M et al. Defective G2 repair in Down syndrome: Effect of caffeine, adenosine and niacinamide in control and X ray irradiated lymphocytes. Clin Genet 1994; 45:25-31.
14. Natarajan AT, Rotteveel AH, van Pieterson et al. Influence of incorporated 5-bromodeoxyuridine on the frequencies of spontaneous and induced sister chromatid exchanges detected by immunological methods. Mutation Res 1986; 163:51-55.
15. Shubber EK, Hamami HA, Allak BMA et al. Sister-chromatid exchanges in lymphocytes from infants with Down's syndrome. Mutation Res 1991; 248:61-72.
16. Finette BA, Rood B, Poseno T et al. Atypical background somatic mutation frequencies in the HPRT locus in children and adults with Down syndrome. Mutation Res 1998; 403:35-43.
17. Yotti LP, Glover TW, Trosko JE et al. Comparative study of X-ray and UV induced cytotoxicity, DNA repair, and mutagenesis in Down syndrome and normal fibroblasts. Pediatr Res 1980; 14:88-92.
18. Natarajan AT, Darroudi F, Mullenders LHF et al. The nature and repair of DNA lesions that lead to chromosomal aberrations induced by ionizing radiations. Mutation Res 1986; 160:231-236.
19. Pfeiffer P, Goedecke W, Obe G. Mechanisms of DNA double strand break repair and their potential to induce chromosomal aberrations. Mutagenesis 2000; 15:289-302.
20. Leonard JC, Merz T. Chromosomal aberrations in irradiated Down's syndrome fibroblasts. Mutation Res 1987; 180:223-230.
21. Athanasiou K, Sideris EG, Bartsocas C. Decreased repair of X-ray induced DNA single strand breaks in lymphocytes in Down's syndrome. Padiatr Res 1980; 14:336-338.
22. Leonard JC, Mertz T, Repair of single strand breaks in normal and trisomic lymphocytes. Mutation Res 1982; 105:417-422.
23. Steiner ME, Woods WG. Normal formation and repair of γ-radiation-induced single and double strand DNA breaks in Down syndrome fibroblasts. Mutation Res 1982; 95:515-523.

24. Chiricolo M, Musa AR, Monti D et al. Enhanced DNA repair in lymphocytes of Down syndrome patients: The influence of zinc nutritional supplementation. Mutation Res 1993; 295:105-111.
25. Countryman PI, Heddle JA, Crawford E. The repair of X-ray induced chromosomal damage in trisomy 21 and normal diploid lymphocytes. Cancer Res 1977; 37:52-58.
26. Preston RJ. X-ray induced chromosome aberrations in down lymphocytes: An explanation for their increased sensitivity. Environ Mutagen 1981; 3:85-89.
27. Lambert B, Hansson K, Bui TH et al. DNA repair and frequency of x-ray and u.v.-light induced chromosome aberrations in leukocytes from patients with Down's syndrome. Ann Hum Genet 1976; 39:293-303.
28. Rebhorn H, Pfeiffenberger H. In vitro life span and "unscheduled DNA synthesis" in subconfluent cultures and clones of trisomic and normal diploid fibroblasts. Mech Ageing Dev 1982; 18:201-208.
29. Ganges MB, Tarone RE, Jiang H et al. Radiosensitive Down syndrome lymphoblastoid lines have normal ionizing-radiation induced inhibition of DNA synthesis. Mutation Res 1988; 194:251-256.
30. Druzhyna N, Nair RG, LeDoux SP et al. Defective repair of oxidative damage in mitochondrial DNA in Down's syndrome. Mutation Res 1998; 409:81-89.
31. Sedl R, Greber S, Schuller E et al. Evidence against increased DNA damage in Down syndrome. Neurosci Lett 1997; 236:137-140.
32. Fang-Kircher SG, Labudova O, Kitzmueller E et al. Increased steady state mRNA levels of DNA-repair genes XRCC1, ERCC2 and ERCC3 in brain of patients with Down syndrome. Life Sci 1999; 64:1689-99.
33. Kola I, Hertzog PJ. Animal models in the study of the biological function of genes on human chromosome 21 and their role in the pathophysiology of Down's syndrome. Hum Mol Genet 1997; 6:1713-1727.
34. Reeves RH, Irving NG, Moran TH et al. A mouse model for Down's syndrome exhibits learning and behaviour deficits. Nat Genet 1995; 11:177-184.
35. Shinohara T, Tmizuka K, Miyabara S et al. Mice containing a human chromosome 21 model behavioral impairment and cardiac anomalies of Down's syndrome. Hum Mol Genet 2001; 10:1163-1175.
36. Roizen NJ, Patterson D. Down's syndrome. Lancet 2003; 361:1281-1289.

CHAPTER 6

The Fanconi Anemia/BRCA Pathway:

FANCD2 at the Crossroad between Repair and Checkpoint Responses to DNA Damage

Massimo Bogliolo and Jordi Surrallés

Abstract

Studies on cancer-prone and rare human genetic disorders often lead to significant advances in our understanding of the complex network of genome stability and DNA repair pathways that have evolved in the human genome to prevent the harmful effects of exposure to DNA damaging agents. One such disorder is Fanconi Anemia, an autosomal recessive disease characterized by an increased spontaneous and DNA cross-linkers induced chromosome instability, progressive pancytopenia and cancer susceptibility. At least eleven genes are involved in Fanconi anemia, including the breast cancer susceptibility gene BRCA2. Six of the Fanconi anemia proteins (FANCA, C, E, F, G and L) assemble in a complex that is required for FANCD2 activation by monoubiquitination in response to DNA damage or during S-phase progression. Active FANCD2 then colocalizes with the product of the breast cancer suscepti- bility gene BRCA1 in discrete nuclear *foci*. FANCD2 is also independently phosphorylated by ATM in response to ionising radiation and interacts with the MRE11/Rad50/NBS1 complex, which is directly involved in homologous recombination DNA repair pathway and in cell cycle checkpoint response to DNA damage. Available data indicate that FANCD2 is involved in cell cycle regulation and DNA repair. Our current knowledge on the functional significance of FA pathway and more specifically FANCD2 and its interacting proteins in pathways of genomic surveillance and maintenance will be discussed in this chapter.

Introduction

In 1967 a Swiss physician, Guido Fanconi, reported his observations of a recessively inherited aplastic anemia in two brothers along with several other physical malformations.[1] Fanconi anemia (FA) is a rare syndrome characterized by progressive pancytopenia, increased spontaneous and mutagen-induced chromosome instability, congenital malformations and can- cer susceptibility. The phenotype of the FA patients reflects that the genes involved in FA are important for chromosome stability, normal embryo development and for the preservation of several types of stem cells. The most frequent birth malformations in FA are growing retardation and abnormalities in the skin (typically cafe-au-lait spots), upper extremities (specially in the thumbs and forearms), kidneys and gastrointestinal system.[2] As 30% of the

DNA Repair and Human Disease, edited by Adayabalam S. Balajee. ©2006 Landes Bioscience and Springer Science+Business Media.

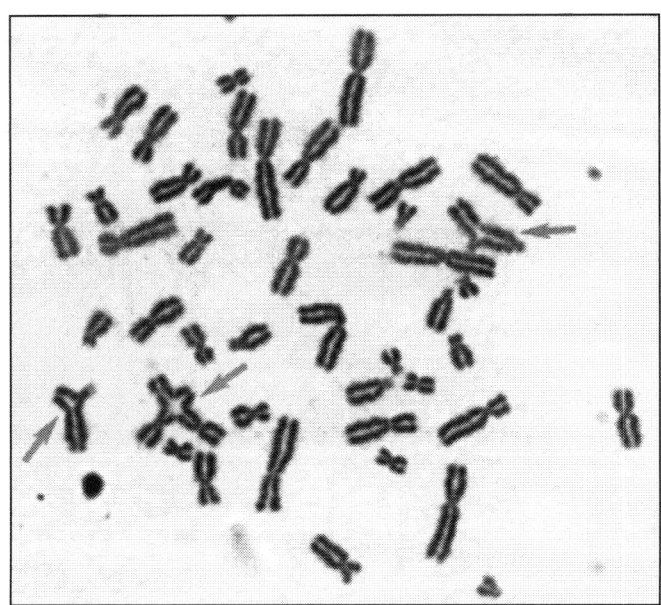

Figure 1. Typical chromatid-type chromosomal aberrations found in FA cells (arrows).

patients have no signs of physical flaws,[3] the ability of DNA cross-linking chemicals such as mitomycin C (MMC) and diepoxybutane (DEB) to induce chromosome aberrations in the patients' blood lymphocytes[2] must be tested for the final confirmation of FA phenotype. The confirmation of FA diagnosis requires cell culturing of peripheral blood lymphocytes, MMC or DEB treatment, analysis of the metaphases and detection of chromosome aberrations, usually of chromatid type such as radial figures (Fig. 1).

FA is a very heterogeneous disease with at least eleven complementation groups described until now[4] (see below). The assignment of a FA patient to a given complementation group is classically resolved by cell fusion-based complementation studies (Fig. 2). The cloning of the genes involved in the most prevalent complementation groups has more recently permitted to develop techniques of genetic subtyping by retroviral transduction of wild type FA genes. Primary blood lymphocytes or cell lines retrovirally transduced with the cDNAs of the various FA genes are tested for the phenotypic correction of MMC hypersensitivity.[5]

FA is usually a fatal disease with a mean survival of twenty-four years.[6] Over 90% of FA patients will most probably develop hematological complications (bone marrow failure, aplastic anemia, thrombocytopenia or pancytopenia) at pediatric age.[6] The cumulative incidence by 40 years of age is 33% for hematologic malignancies and 28% for solid tumors, with a strong predisposition to squamous cells carcinoma, especially of the head and neck region. In fact, there is a 500-fold increase in the cumulative incidence of head and neck cancer in FA patients compared with the normal population.[6]

The phenotype of primary or immortalized FA-derived cell lines is characterized by their extreme sensitivity to agents that produce DNA interstrand crosslinks such as MMC, DEB and cisplatin, resembling the phenotype of the patients' blood lymphocytes. After treatment with these agents, FA cell lines show increased chromosome instability. Another feature observed in FA cell lines is a delay in G2/M or late S-phase of the cell cycle. This delay occurs spontaneously and increases greatly after treatment of the cells with crosslinking agents.[7]

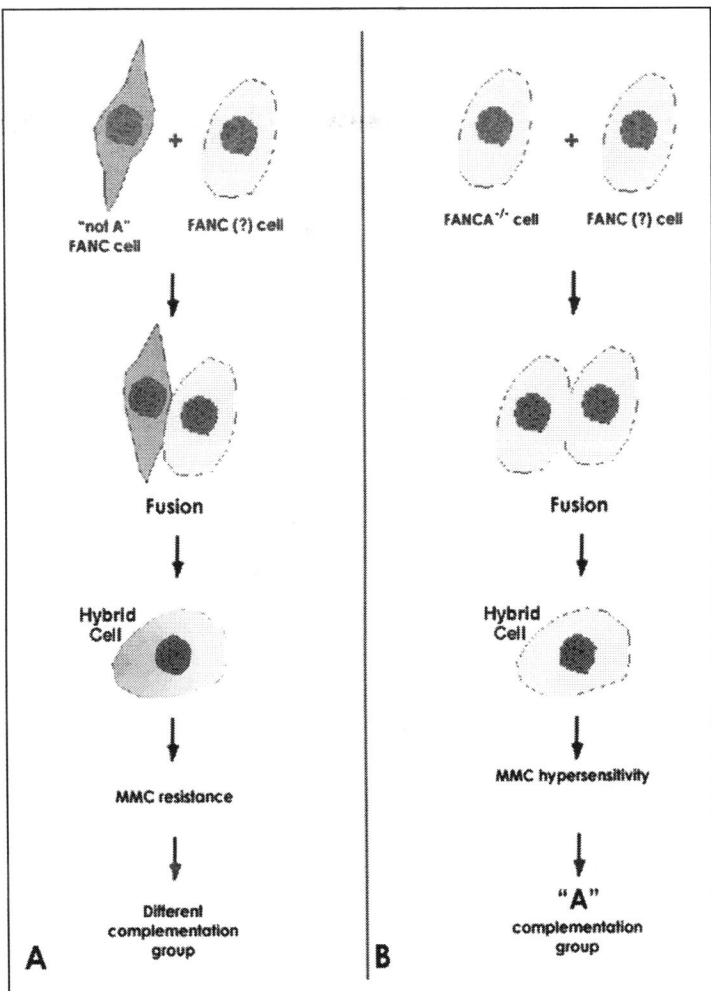

Figure 2. Complementation group determination in FA by cell fusion. A) When cells derived from a patient of an unknown complementation group (yellow) are fused with cells of a known but different complementation group (blue) the FA phenotype is functionally complemented and the hybrid cell is resistant to MMC indicating that this patient belongs to a different complementation group. B) When cells of unknown complementation group are fused with cells of the same complementation group, the resulting hybrid cells are still hypersensitive to MMC indicating that the two cells lines have mutations in the same gene and, consequently, they do not complement each other. A color version of this figure is available online at http://www.Eurekah.com.

The Genetics of Fanconi Anemia

FA is a very rare autosomal recessive genetic disease with a prevalence of 1-5 per million.[2] The frequency of heterozygote carriers in the population is estimated to vary between 1:200 to 1:300. Some ethnic groups have a higher prevalence of FA due to a founder effect and consanguinity. Two examples are the Afrikaner population of South Africa and the Ashkenazi Jewish with a carrier frequency of 1/77 and 1/90, respectively. An even higher prevalence of FA is found in Spanish Gypsies (Callen et al, manuscript submitted). It is known that a common mutation in the FANCC gene is responsible for the majority of FA patients in Ashkenazi Jews[8] and similarly the majority of the patients of the Afrikaner population of South Africa have an identical mutation in FANCA. Molecular and genealogical evidences obtained in this later group confirmed the existence of a founder effect for FA in South Africa.[9]

Cell fusion complementation studies (Fig. 2) revealed the existence of at least eleven FA complementation groups: FANCA, FANCB, FANCC, BRCA2/FANCD1, FANCD2, FANCE, FANCF, FANCG, FANCL, FANCI and FANCJ. Most of the FA genes, with the exception of FANCB, FANCI and FANCJ, have been cloned and characterized.[4,10,11] According to the International FA Register, complementation groups A (65%), C (15%), and G (10%) account for at least 90% of the FA patients in a population of 755 FA patients from North America while the rest of the subtypes are relatively rare.[4,6] Cell lines derived from FANCD1 patients have biallelic mutations in the breast cancer susceptibility gene BRCA2 and express truncated BRCA2 proteins.[12] The BRCA2 protein participates in DNA repair by homologous recombination.[13-15] Mutations in a DNA repair gene, BRCA2, in cells derived from FANCD1 patients strongly suggest that at least a subset of FA phenotype is causally linked to DNA repair deficiency. Recently, afflicted patients in four FA families were found to be compound heterozygotes for BRCA2 mutations. The individuals affected, exhibit the classical FA phenotype (higher levels of spontaneous and chemically induced chromosome aberrations and hematological complications) together with brain tumour development. In one of the kindreds, a family history for breast cancer was also observed. The cooccurrence of FA phenotype, brain tumours and breast cancer constitutes a new syndromic association and highlights the critical link between FA pathway and cancer development.[16]

FANCD2 as a Key Player in Fanconi Anemia

The molecular biology of FA is steadily becoming clearer and now it is well established that most of the FA proteins (A, C, E, F, G, L and perhaps I) assemble in a nuclear complex required for the activation of FANCD2 via monoubiquitination at lysine 561.[10,11,17] Monoubiquitinated FANCD2 (FANCD2-L) is bigger than its inactive isoform, (FANCD2-S) which can be easily distinguished by western blot analysis. When one of the genes in the FA nuclear complex is mutated, this complex is not properly formed resulting in a single FANCD2-S band in a western blot (Fig. 3). An exception to this is FANCI, which although dispensable for FA core complex formation as judged by coimmunoprecipitation studies, is required for FANCD2 monoubiquitination indicating that FANCI is upstream of FANCD2. In contrast, FANCJ and FANCD1/BRCA2 deficient cells are able to form a FA multiprotein core complex and to monoubiquitinate FANCD2, suggesting that their defect in the FA pathway is downstream to the crucial step of FANCD2 activation.[4] The importance of FANCD2 as one of the key players of FA pathway is illustrated by the evolutionary conservation of FANCD2 together with FANCL[11] in distant species such as *Drosophila melanogaster, Caenorabditis elegans, Arabidopsis thaliana*.[18] Cloning and sequencing of the Drosophila FANCD2 gene, for instance, revealed a remarkable functional conservation of important features, such as the residue K561, during evolution[19] (Fig. 4).

The monoubiquitinated isoform of FANCD2 associates with the repair protein BRCA1 in DNA damage-induced nuclear *foci*.[17] This *foci* formation is induced not only by crosslinking agents but also by other DNA damaging agents such as UVC light and ionising radiation.[17] This evidence suggests a wider field of action for FANCD2 probably in cooperation with other DNA damage response proteins known to also colocalize in nuclear *foci* with BRCA1 after DNA damage.[20] In addition, FANCD2 undergoes monoubiquitination during the S-phase of the cell cycle even in the absence of mutagen induced-DNA damage and the monoubiquitinated FANCD2 colocalizes with BRCA1 and RAD51 in S-phase-specific nuclear *foci*.[21]

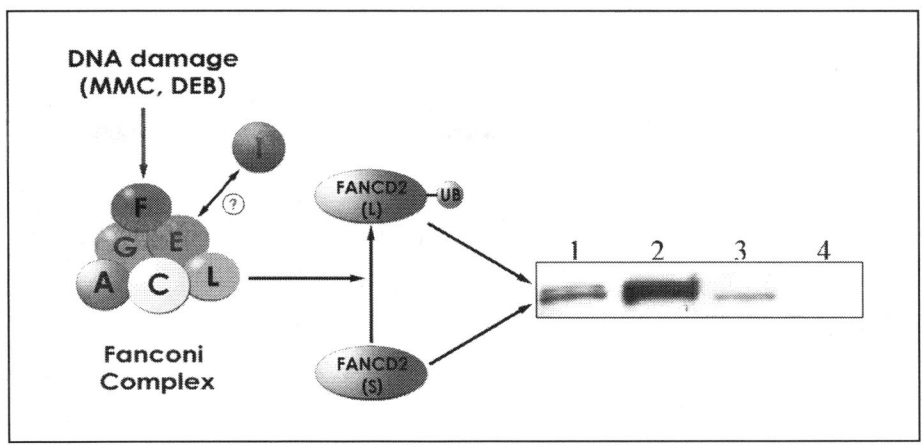

Figure 3. FA nuclear complex and activation of FANCD2. The FA nuclear complex mediates the activation by monoubiquitination of FANCD2 and western blot with antibodies against FANCD2 reveals two bands (FANCD2-L and S) in normal cells (lane 1), one FANCD2-S (short) band in FA-A cells (lane 3) and no signal in the FANCD2 $^{-/-}$ reference cell line PD20 (lane 4) indicating the absence of this protein. Gene transfer of wild-type FANCD2 cDNA in the PD20 cell line results in FANCD2 (over) expression and phenotypic reversion (lane 2).

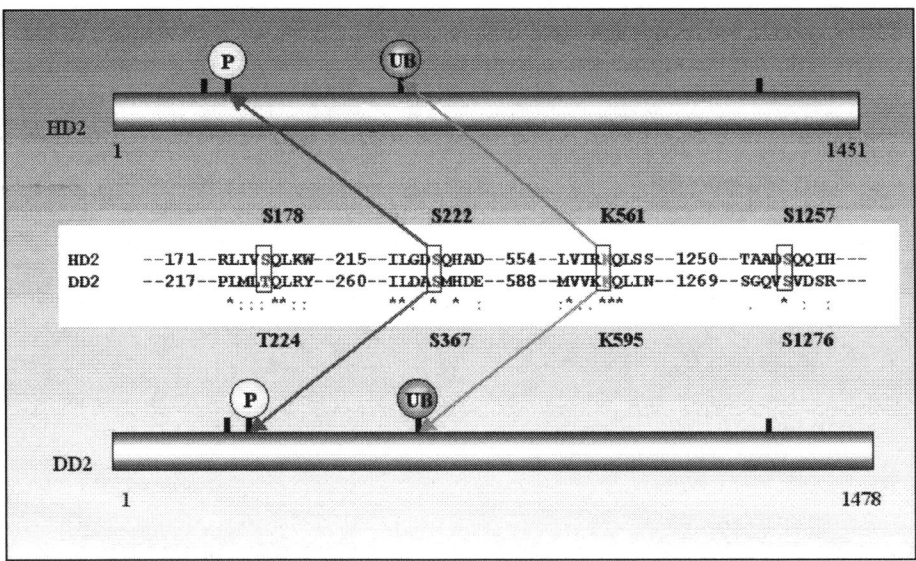

Figure 4. Human and *Drosophila* FANCD2 post-translational modification sites. Sequence and structural biocomputational analysis indicated a 23% of identity and 43% of similarity between human (HD2) and the *Drosophila* (DD2) FANCD2 proteins. Human ATM phosphorylation site at S222 and the complex-dependent monoubiquitination site at K561 are highly conserved in Drosophila at positions S267 and K595, respectively. Other putative ATM-dependent phosphorylation sites in the human protein (S178 and S1257) are also conserved in Drosophila at positions T224 and S1276, respectively.

Recent studies have shown that ubiquitin ligase activity of BRCA1[22] is not directly responsible of FANCD2 monoubiquitination, since this post-translational modification also occurs in BRCA1 deficient cells.[23] The main catalytic subunit involved in FANCD2 monoubiquitination is the gene involved in the complementation group FA-L (FANCL) known as ubiquitin ligase PHF9. FANCL-PHF9 possesses E3 ubiquitin ligase activity in vitro and it has a crucial role in the FA pathway.[11]

In addition to monoubiquitination, a second posttranslational modification of FANCD2 also seems to be critical for DNA damage response. ATM, the product of the gene mutated in the chromosome fragility syndrome ataxia telangiectasia, phosphorylates FANCD2 at serine 222 in response to ionising radiation and activates G1-S checkpoint control.[24] Ser 222 is also highly conserved in evolution.[19] The ATM gene encodes for a protein kinase triggered by ionising radiation that phosphorylates a number of downstream targets such as the tumour suppressor gene p53, the Nijmegen breakage syndrome (NBS) protein NBS1, CHK2 and BRCA1, all of which are involved in the S-phase checkpoint activated by ionising radiation. Biallelic loss of ATM gene or of one of its substrates produces a defect in this S-phase checkpoint characterized by a radioresistant DNA synthesis after exposure to ionising radiation.[25] Among the FA cell lines, only FANCD2 are mildly sensitive to ionising radiation and are defective in the S-phase checkpoint induced by ionising radiation.[24] Thus, it appears that the two post-translational modifications of FANCD2 are important for DNA repair and intra-S-phase checkpoint regulation after DNA damage. The two posttranslational modifications are independent: disruption of the FA complex results in the lack of the activation of FANCD2 by monoubiquitination and in hypersensitivity to cross-linking agents, without affecting the S-phase checkpoint. On the other hand, the lack of ATM-dependent FANCD2 phosphorylation leads to an inactivation of an S-phase cell cycle checkpoint and to radioresistant DNA synthesis, but does not affect FANCD2 monoubiquitination by the FA complex after cross-linkers induced DNA damage. Further supporting this hypothesis is the fact that ATM cell lines are not hypersensitive to cross-linkers and all the FA cells with the only exception of FANCD2, are not hypersensitive to ionising radiations. These observation suggest that FANCD2 could be where the FA and ATM signaling pathways converge.[24] ATM-dependent response to ionising radiations is not only mediated through the interaction with FANCD2 but also by NBS1 which is phosphorylated by ATM. After phosphorylation, NBS1 interacts with FANCD2 promoting its ATM-mediated phosphorylation and activating the S-phase cell cycle checkpoint and radioresistant DNA synthesis. Accordingly, ATM-dependent phosphorylation of FANCD2 is defective in NBS1 deficient cells (Fig. 6B).[26]

High FANCD2 expression was found in human germ cells and in haematopoietic system and other highly proliferative tissues known to be predisposed to cancer development in FA patients, particularly in the head and neck region and in the uterine cervix. These observations point to an important role for FANCD2 in the maintenance of genomic integrity during cellular proliferation within these tissues.[27]

DNA Repair and Cell Cycle Regulation Downstream FANCD2

The phenotype of the FA patients tells us how the FA pathway is important in genome stability, embryonic development and in the maintenance of several types of stem cells. Cultured cells from FA patients display a high level of spontaneous chromosome breaks and an increased frequency of intragenic deletions, suggesting that FA cells may have deficiencies in the repair of DNA double strand breaks in addition to DNA cross links. Eukaryotic cells possess two modes of repair pathways for dealing with double strand breaks: (1) nonhomologous end-joining pathway (NHEJ) and (2) homologous recombination (HR) repair. Among these two, NHEJ is relatively more simple where the DNA bearing a double strand break is simply ligated back.[28] It has been observed that the FA cell lines are impaired in the DNA repair by

NHEJ as they show abnormal rearrangements associated with VDJ recombination[29] and impaired fidelity in blunt DNA end joining.[30,31] A deficient DNA end joining activity in extracts from FA fibroblasts has also been reported. The Fanconi anemia extracts had 3- to 9-fold less DNA end joining activity and rejoined substrates with significantly less fidelity than normal extracts. Protein expression levels of the DNA-dependent protein kinase (DNA-PK)/ Ku-dependent NHEJ proteins Xrcc4, DNA ligase IV, Ku70, and Ku86 in FA and normal extracts were indistinguishable. Taken together, these results suggest that the FA fibroblast extracts have a deficiency in a DNA end joining process that is distinct from the DNA-PK/ Ku-dependent NHEJ pathway.[32]

The HR repair is an error-free DNA repair system specific of S and G2-phases of the cell cycle, when two chromatids are present. In this repair process, the undamaged chromatid is used as a template for the error free exchange and repair of the other chromatid (Fig. 5). As mentioned earlier, cell lines derived from FANCD1 patients have biallelic mutations in BRCA2, and FANCD1 cells express truncated BRCA2 proteins.[12] This finding is particularly relevant from the DNA repair point of view because several studies have implicated BRCA2 in HR repair during S-phase.[13-15] These observations convincingly demonstrate that a potential DNA repair gene, BRCA2, is also implicated in the pathogenesis of FA. Research on the roles of BRCA1 and BRCA2 led to the finding that both proteins interact with the recombinase Rad51 and all of them are involved in HR pathway.[33] The findings that BRCA2/FANCD1 and Rad51 are downstream in the FA/BRCA pathway and that FANCD2 forms nuclear *foci* with BRCA1 and Rad51 during S-phase, raise the possibility that the FA/BRCA pathway play a role in HR repair of interstrand crosslinks and double strand breaks during S-phase.[17,21] Cells impaired in HR repair such as BRCA1 and BRCA2 deficient cells are very sensitive to DNA crosslinking agents indicating that HR is important in the processing of DNA crosslinks.[34] Functional linkage between the FA pathway and the breast cancer susceptibility genes has been demonstrated by the interaction of FANCD2 with BRCA1[17] and by the discovery that the FANCD1 gene is identical to BRCA2.[12] Also other FA proteins interact with BRCA genes: an interaction between FANCA and BRCA1[35] and binding of FANCG to two separate sites of BRCA2, located on either side of the BRC repeats have been demonstrated using yeast two-hybrid system. Furthermore, FANCG can be coimmunoprecipitated with BRCA2 from human cell extracts and FANCG colocalizes in the nuclear *foci* with both BRCA2 and RAD51 following MMC-induced DNA damage.[36] A role for FA proteins in DNA repair is further strengthened by their association with BLM helicase whose unwinding activity is critical for DNA repair activities. A complex isolated from HeLa cell extract using antibody to Bloom syndrome protein (BLM) has been shown to contain at least five of the FA complementation group proteins (A, C, G, E and F). This complex exhibits a DNA-unwinding activity, predominantly owing to the presence of BLM helicase because such a complex isolated from BLM-deficient cells lack such an activity. These results have been confirmed by the fact that the same complex could be isolated with an antibody against FANCA. It is interesting to note that cells from Bloom syndrome patients are characterized by spontaneous and UV induced elevation of sister chromatid exchanges which are attributed to a deficiency in HR pathway. Coimmunoprecipitation of FA proteins with BLM in the same complex suggests that both BLM and FA proteins may operate in the same repair pathway.[37] As previously mentioned, ATM directly phosphorylates FANCD2 on Ser 222 which is required for the activation of S-phase checkpoint.[24] ATM not only phosphorylates FANCD2 but also an increasing number of proteins involved in both HR and NHEJ.[38] NBS1 is phosphorylated by ATM on Ser 343 and forms a complex with MRE11 and Rad50 (RMN complex). The RMN complex is a very important regulator of cell cycle checkpoint and DNA repair responses in all eukaryotic cells and the complex is directly involved in HR.[39-41] A subset of NBS patients shares symptoms with FA patients such as bone marrow failure[42] and some NBS cell lines shows a mild sensitivity to MMC.[26] The identification of a patient with an

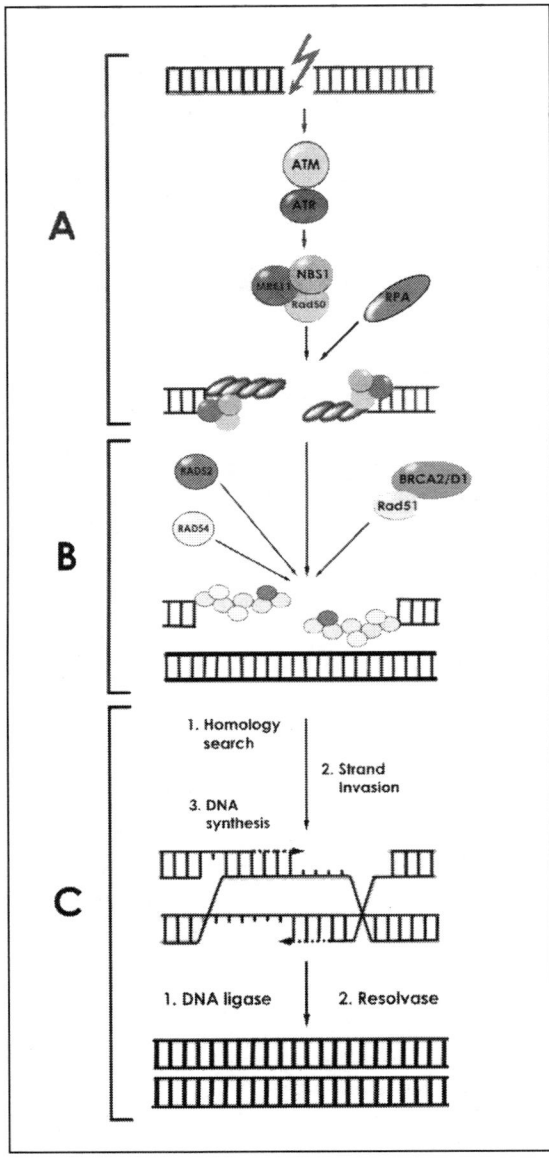

Figure 5. Schematic representation of homologous-recombination repair during S-phase. DNA damage activates ATM/ATR that triggers by phosphorylation the RAD50/MRE11/NBS1 complex. The 5'-3' exonuclease activity of this complex exposes both 3' ends (A). To promote strand invasion into homologous sequences, RPA facilitates assembly of a RAD51 nucleoprotein filament and RAD52 stimulates filament assembly. BRCA2/FANCD1 is essential for loading Rad51 to the single stranded DNA, explaining why mutation-induced Rad51 foci are not formed in FANCD1 deficient cells.[71] RAD51 has the ability to exchange the single strand with the same sequence from a double-stranded DNA molecule. RAD54, a member of the SWI/SNF family of DNA-dependent ATPases, is a candidate for the complex chromatin transactions associated with HR (B). After identification of the identical sister chromatid sequence, the intact double-stranded copy is used as a template to properly heal the broken ends by DNA synthesis. Finally, the so-called Holliday-junctions are resolved by resolvases (C).

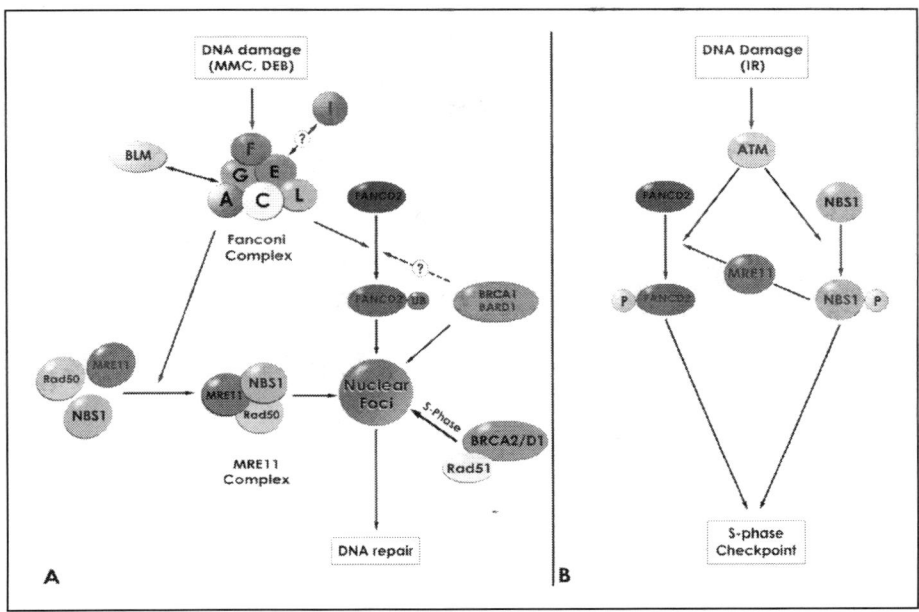

Figure 6. FANCD2 is a central node between repair and cell cycle response to DNA damage. A) In response to DNA damage induced by DNA crosslinkers, the FA nuclear complex activates FANCD2 monoubiquitination and, via FANCC, the RMN complex assembly. FANCD2, the RMN complex, BRCA1/2 and Rad51 colocalize in nuclear foci probably involved in DNA repair. B) In response to ionising radiation ATM triggers the S-phase checkpoint via phosphorylation of NBS1 and subsequently of FANCD2. The ATM-dependent phosphorylation of FANCD2 in response to ionising radiation also requires a functional RMN complex.

atypical FA phenotype bearing biallelic mutations in NBS1 gene conclusively pointed out an interaction between FA and NBS.[26] In support, NBS1 and the monoubiquitinated isoform of FANCD2 colocalize in sub-nuclear *foci* in response to DNA damage by crosslinking agents. The disruption of the monoubiquitination site on FANCD2 or the disruption of the MRE11-interacting carboxyl terminus of NBS1 prevents these *foci* to form, resulting in MMC hypersensitivity.[26] Hence the FA pathway and the RMN complex must cooperate in the cellular response to DNA damage by crosslinking agents (Fig. 6A). After ionising radiation treatment, NBS1 is phosphorylated by ATM and the phosphorylated form of NBS1 promotes the subsequent phosphorylation of FANCD2 by ATM, since ATM-dependent phosphorylation of FANCD2 is defective in Nbs1 cells (Fig. 6B) Therefore, FANCD2 and the RMN complex also cooperate in the checkpoint response after DNA damage by ionising radiation. All these observations indicate that the FA pathway in general and FANCD2 in particular function at the intersection of two inter-dependent DNA repair and cell cycle signaling pathways initiated by two DNA damaging agents. In response to MMC and other crosslinking agents, active FANCD2 assembles with BRCA1, MRE11 complex, BRCA2 and Rad51 in nuclear *foci* during HR (Fig. 6A). In case of radiation treatment, ATM phosphorylates FANCD2 and NBS1 thereby imposing the intra-S-phase checkpoint regulation[26] (Fig. 6B). The interaction with the RMN complex is not only a prerogative for FANCD2, since RMN assembly is defective in FANCC cells in response to MMC. These alterations in the assembly of DNA-repair proteins and cell cycle regulators in FA can explain the DNA-damage processing anomalies observed in FA cells and for the genetic instability and the cancer predisposition of this syndrome[43] (Fig. 6A).

Mouse Models of FA

Expression of the mouse homologs in human FA cells of the same complementation group corrects the hypersensitivity to DNA-crosslinking agents, suggestive of a functional conservation in FA proteins between humans and mice.[44] Studies using these animal models provide valuable insights into the pathways involved in FA, and are also useful for testing novel genetic treatments for FA.[45] Knockout mice for the genes involved in the A,[46] C,[47] G[48] and D2[49] and D1/BRCA2[50] FA complementation groups are currently available. Except for the FANCD1 and FANCD2 deficient mice, all other null mice share an almost undistinguishable pheno-type, supporting the hypothesis that the FANCA, C and G proteins share a common function in the FA pathway.[44,45] This is further supported by the fact that FANCA/FANCC double mutants have the same phenotype of the single mutants.[51] FA phenotype in mice is far milder than in human.[45] FA A, C and G mouse models have normal lifespan, no developmental abnormalities, except a decrease in bodyweight and in the size of testes and ovaries. In contrast to FA patients, FA null mice do not exhibit anemia, with the exception of a modest thromb-ocytopenia,[45] but not increased tumour development.[44,45] FA knockout mice show reduced fertility, a feature consistent with the FA patients.[52,53] However, when FANCC mice are subjected to MMC treatment at a concentration that does not affect the normal mice, there is a progressive decrease in all peripheral blood parameters and bone marrow failure, causing death within 3-8 weeks.[54] In addition, a marked in vitro growth defect was observed in FANCA$^{-/-}$ hematopoietic progenitors after in vitro stimulation.[45] Hypersensitivity to cytokines has also been described showing how the haematopoietic cells of FA mice are more susceptible to apoptosis[55] and impaired in their ability to repopulate the myeloid and lymphoid lineage.[56] Cells cultured from all the FA mouse models show increased chromosomal aberrations when exposed to DNA cross-linking agents,[54,57] confirming that the FA pathway in mice seems to be identical to that of humans. Therefore in mouse models it appears that the loss of FA genes, whose protein products are involved in the formation of the FA nuclear complex, does not endanger survival under normal circumstances, but impairs the ability to respond to environmental insults.[44]

BRCA2 null mice are not viable[58] but if the 3' region of BRCA2 gene is disrupted, then the mice are viable, probably because the truncated protein still conserves some partial activity.[50] The phenotype of these mice bearing sub-lethal mutation of *BRCA2* is peculiar: they do not develop breast cancer but they are of small size, have skeletal defects, hypogonadism, chromosome instability, cancer predisposition and MMC hypersensitivity.[50,59] FANCD1$^{-/-}$ human cell lines have mutations exactly in the carboxyl terminal of the BRCA2 protein[12] and, consequently these sub lethal BRCA2 null mice could be a considered by all means a mouse model for FANCD1 patients.

Similar to human FA patients and other FA mouse models, FANCD2 mutant mice are hypersensitive to DNA interstrand cross-links and show loss of germinal cells. Furthermore, a higher frequency of chromosomal mispairing in male meiosis is observed as compared to the wild type mice. FANCD2 mutant mice also exhibit characteristics not observed in other mice with disruptions of other FA genes, except for BRCA2 knockout mice. These include micropthalmia, perinatal lethality, and development of malignancies in different epithelial cells.[49] The increased tumour formation in FANCD2 mutant mice shows that only when the function of FANCD2 is completely lost there is also cancer proneness, emphasizing again the crucial role of FANCD2 in the FA pathway. Actually, the phenotype of FANCD2 null mice is very similar to the phenotype of FANCD1/BRCA2 knockout mice and the tumors spectrum in the two mouse models is comparable with a similar predisposition to epithelial cancer.[59] This phenotypic overlap is consistent with a common function for both proteins in the same pathway, regulating genomic stability.

The FA Pathway in Meiosis and Chromatin Remodeling

BRCA1 can be detected at the human synaptonemal complexes in human zygotene and pachytene spermatocytes implying a functional role for BRCA1 in the meiotic and mitotic cell cycles and in the control of recombination and genome integrity.[60] The association of FANCD2 and BRCA1 in mitotic cells suggested that FANCD2 might also colocalize with BRCA1 during meiosis at the same level. Analysis of mouse spermatocytes with antibodies against FANCD2 revealed an intense staining at the level of the impaired axes of the sex chromosomes during pachynema and diplonema. This intense staining colocalizes with BRCA1 staining.[17] This data, together with the phenotype of fertility defects in both humans and mouse models, suggest that FANCD2 may be required for the chromosome segregation during meiosis in male spermatocytes.

Chromatin remodeling and transcription regulatory functions have been described for BRCA1 and BRCA2, and more specifically for the C-terminal domain of BRCA1. The C-terminal 20 aminoacids of FANCD2 contains a highly acidic domain similarly to the non-histone chromatin high-mobility group proteins (HMG) suggesting a possible mechanism for its chromatin association.[17] The biological significance of this domain is, however, unclear as it is not conserved in Drosophila.[19] Other chromatin modifier factors interacting with FA proteins are BRG1, as a subunit of the SWI/SNF complex.[61] BRCA1 is also associated with a human SWI/SNF-related complex, linking chromatin remodeling to breast cancer[62] and BRCA2 is a histone acetyl transferase.[63] A very recent report using the yeast two hybrid system shows that the FA proteins FANCA, FANCC and FANCG interact with many nuclear and cytoplasmatic proteins including proteins involved in transcription regulation, signaling, oxidative metabolism, and intracellular transport, thereby strongly supporting the hypothesis that the FA proteins are functionally involved in a multitude of cellular routes.[64]

It has been proposed that FANCC may be related to a transcriptional repression pathway involved in chromatin remodeling through interaction with the Fanconi Anemia Zinc finger (FAZF) protein[65] and it is also known that the FA proteins bind to chromatin and nuclear matrix but are excluded from condensed mitotic chromosomes.[66] In a recent study we reported that, unlike the overall genome, the sensitivity of chromosome 1 constitutive heterochromatin to the chromosome breaking activity of cross-linking agents is independent of a functional FA pathway, indicating that the action of the FA pathway is nonrandomly distributed throughout the human genome.[67] These findings would connect the action of the FA proteins with chromatin remodeling and transcriptional factors. Taken together all these cumulative data are indicative of a role for FA pathway in transcriptional activity and chromatin remodeling or vice versa. There are many instances where transcription and chromatin remodeling play important modulatory roles in other DNA repair systems such as nucleotide excision repair leading to clustered transcription coupled repair activity in the human genome.[68] Since transcription takes place in the same substrate as repair, replication and recombination, it is therefore not surprising that these processes are physically and functionally connected.[69] Future experiments will probably uncover the functional significance of the crosstalk between the FA pathway, chromatin remodeling and DNA repair/transcriptional factors.

Concluding Remarks

The main conclusion that arises from this chapter is that the FA pathway, and especially FANCD2, is a central player in a network of DNA repair and checkpoint responses to DNA damage, but several specific questions remain unanswered. What is the exact molecular nature of the relationship between the FA pathway and HR repair? There are numerous indications and evidences of this interaction but what exactly the FA pathway and FANCD2 do remains unknown. What is the role of FA pathway in the other phases of the cell cycle other than

S-phase? In the last years, the majority of studies have focused on the role of the FA pathway in S-phase and in HR, but we must not forget that this is only the tip of the iceberg. MMC exerts its DNA damaging action only in S-phase, but the FA pathway is activated by a wide array of chemical and physical agents—such as UVC and ionising radiation for instance[17]—that produce damage repaired by mechanisms (nucleotide excision repair, base excision repair, NHEJ) that operate not only in S-phase but also in G1 and G2 phases. Additionally, we must bear in mind that a defect in any of the FA proteins has a profound impact on so many fundamental cellular activities that trying to explain all of these outcomes only on the basis of a deficiency in FANCD2 activation, is probably too simplistic, if not misleading at all. There are an amazing number of studies focused on the many additional roles that the other FA proteins undertake both in the cytoplasm and in the nucleus, such as the processing of reactive oxygen species, the regulation of apoptosis and the maintenance of telomere integrity.[10,70] Even so the FA/BRCA pathway[71] can be considered as the most exciting discovery of the last few years in the field of FA, and clarifying the many twists and turns of this pathway would prove to be an exciting scientific challenge for the years to come.

Acknowledgements

FA research in our Group is in part supported by the Generalitat de Catalunya (project SGR-00197-2002), the Spanish Ministry of Health and Consumption (projects FIS PI020145 and FIS-Red G03/073), the Spanish Ministry of Science and Technology (projects SAF 2002-03234, SAF2002-11833-E and SAF 2003-00328) and the Commission of the European Union (projects FIGH-CT-2002-00217, FI6R-CT-2003-508842 and HPMF-CT-2001-01330). M.B. is supported by a long-term postdoctoral Marie Curie fellowship awarded by the Commission of the European Union. J.S. is supported by a "Ramón y Cajal" project entitled "Genome stability and DNA repair" cofinanced by the Spanish Ministry of Science and Technology and the Universitat Autònoma de Barcelona.

References

1. Fanconi G. Familial constitutional panmyelocytopathy, Fanconi anemia (F.A.). I. Clinical aspects. Semin Hematol 1967; 4:233-240.
2. Joenje H, Patel KJ. The emerging genetic and molecular basis of Fanconi anaemia. Nat Rev Genet 2001; 2:446-457.
3. Strathdee CA, Buchwald M. Molecular and cellular biology of Fanconi anaemia. Am J Pediatric Hematol Oncol 1992; 14:177-185.
4. Levitus M, Rooimans MA, Steltenpool J et al. Heterogeneity in Fanconi anemia: Evidence for two new genetic subtypes. Blood 2003, Epub ahead of print.
5. Hanenberg H, Batish SD, Pollok KE et al. Phenotypic correction of primary Fanconi anemia T cells with retroviral vectors as a diagnostic tool. Exp Hematol 2002; 30:410-420.
6. Kutler DI, Singh B, Satagopan J et al. A 20-year perspective on the International Fanconi Anaemia Registry (IFAR). Blood 2003; 101:1249-1256.
7. Kaiser TN, Lojewski A, Dougherty C et al. Flow cytometric characterization of the response of Fanconi's anemia cells to mitomycin C treatment. Cytometry 1982; 2:291-297.
8. Whitney MA, Saito H, Jakobs PM et al. A common mutation in the FACC gene causes Fanconi anaemia in Ashkenazi Jews. Nat Genet 1993; 4:202-205.
9. Tipping AJ, Pearson T, Morgan NV et al. Molecular and genealogical evidence for a founder effect in Fanconi anemia families of the Afrikaner population of South Africa. Proc Natl Acad Sci USA 2001; 98:5734-5739.
10. Bogliolo M, Cabre O, Callen E et al. The fanconi anaemia genome stability and tumour suppressor network. Mutagenesis 2002; 17:529-538.
11. Meetei AR, de Winter JP, Medhurst AL et al. A novel ubiquitin ligase is deficient in Fanconi anemia. Nat Genet 2003; 35:165-170.

12. Howlett NG, Taniguchi T, Olson S et al. Biallelic inactivation of BRCA2 in Fanconi anemia. Science 2002; 297:606-609.

13. Davies AA, Masson JY, McIlwraith MJ et al. Role of BRCA2 in control of the RAD51 recombination and DNA repair protein. Mol Cell 2001; 7:273-282.

14. Scully R, Puget N, Vlasakova K. DNA polymerase stalling, sister chromatid recombination and the BRCA genes. Oncogene 2000; 19:6176-6183.

15. Moynahan ME, Pierce AJ, Jasin M. BRCA2 is required for homology-directed repair of chromosomal breaks. Mol Cell 2001; 7:263-272.

16. Offit K, Levran O, Mullaney B et al. Shared genetic susceptibility to breast cancer, brain tumors, and Fanconi anemia. J Natl Cancer Inst 2003; 95:1548-1551.

17. Garcia-Higuera I, Taniguchi T, Ganesan S et al. Interaction of the Fanconi Anemia Proteins and BRCA1 in a Common Pathway. Mol Cell 2001; 7:249-262.

18. Timmers C, Taniguchi T, Hejna J et al. Positional cloning of a novel Fanconi anemia gene, FANCD2. Mol Cell 2001; 7:241-248.

19. Castillo V, Cabre O, Marcos R et al. Molecular cloning of the Drosophila Fanconi anaemia gene FANCD2 cDNA. DNA Repair (Amst) 2003; 2:751-758.

20. Futaki M, Liu JM. Chromosomal breakage syndromes and the BRCA1 genome surveillance complex. Trends Mol Med 2001; 7:560-565.

21. Taniguchi T, Garcia-Higuera I, Xu B et al. Convergence of the Fanconi anaemia and Ataxia Telangiectasia signaling pathways. Cell 2002; 109:459-472.

22. Taniguchi T, Garcia-Higuera I, Andreassen PR et al. S-phase specific interaction of the Fanconi anemia protein, FANCD2, with BRCA1 and RAD51. Blood 2002; 100:2414-2420.

23. Grompe M, D'Andrea AD. Fanconi anemia and DNA repair. Hum Mol Genet 2001; 10:2253-2259.

24. Vandenberg CJ, Gergely F, Ong CY et al. BRCA1-independent ubiquitination of FANCD2. Mol Cell 2003; 12:247-254.

25. Taniguchi T, Garcia-Higuera I, Xu B et al. Convergence of the fanconi anemia and ataxia telangiectasia signaling pathways. Cell 2002; 109:459-472.

26. Shiloh Y. ATM and related protein kinases: Safeguarding genome integrity. Nat Rev Cancer 2003; 3:155-168.

27. Nakanishi K, Taniguchi T, Ranganathan V et al. Interaction of FANCD2 and NBS1 in the DNA damage response. Nat Cell Biol 2002; 4:913-920.

28. Holzel M, van Diest PJ, Bier P et al. FANCD2 protein is expressed in proliferating cells of human tissues that are cancer-prone in Fanconi anaemia. J Pathol 2003; 201:198-203.

29. Lieber MR, Ma Y, Pannicke U et al. Mechanism and regulation of human nonhomologous DNA end-joining. Nat Rev Mol Cell Biol 2003; 4:712-720.

30. Smith J, Andrau JC, Kallenbach S et al. Abnormal rearrangements associated with V(D)J recombination in Fanconi anemia. J Mol Biol 1998; 281:815-825.

31. Escarceller M, Rousset S, Moustacchi E et al. The fidelity of double strand breaks processing is impaired in complementation groups B and D of Fanconi anemia, a genetic instability syndrome. Somat Cell Mol Genet 1997; 23:401-411.

32. Escarceller M, Buchwald M, Singleton BK et al. Fanconi anemia C gene product plays a role in the fidelity of blunt DNA end-joining. J Mol Biol 1998; 279:375-385.

33. Lundberg R, Mavinakere M, Campbell C. Deficient DNA end joining activity in extracts from fanconi anemia fibroblasts. J Biol Chem 2001; 276:9543-9549.

34. Tutt A, Ashworth A. The relationship between the roles of BRCA genes in DNA repair and cancer predisposition. Trends Mol Med 2002; 8:571-576.

35. Dronkert ML, Kanaar R. Repair of DNA interstrand cross-links. Mutat Res 2001; 486:217-247.

36. Folias A, Matkovic M, Bruun D et al. BRCA1 interacts directly with the Fanconi anaemia protein FANCA. Hum Mol Genet 2002; 11:2591-2597.

37. Hussain S, Witt E, Huber PA et al. Direct interaction of the Fanconi anaemia protein FANCG with BRCA2/FANCD1. Hum Mol Genet 2003; 12:2503-2510.

38. Meetei AR, Sechi S, Wallisch M et al. A multiprotein nuclear complex connects Fanconi anemia and Bloom syndrome. Mol Cell Biol 2003; 23:3417-3426.

39. Kastan MB, Lim DS, Kim ST et al. Multiple signalling pathways involving ATM. Cold Spring Harb Symp Quant Biol 2000; 65:521-526.

40. Tauchi H, Matsuura S, Kobayashi J et al. Nijmegen breakage syndrome gene, NBS1, and molecular links to factors for genome stability. Oncogene 2002; 21:8967-8980.

41. Thompson LH, Schild D. Recombinational DNA repair and human disease. Mutat Res 2002; 509:49-78.

42. Petrini JH. The Mre11 complex and ATM: Collaborating to navigate S phase. Curr Opin Cell Biol 2000; 12:293-296.

43. Resnick IB, Kondratenko I, Togoev O et al. Nijmegen breakage syndrome: Clinical characteristics and mutation analysis in eight unrelated Russian families. J Pediatr 2002; 140:355-361.

44. Pichierri P, Averbeck D, Rosselli F. DNA cross-link-dependent RAD50/MRE11/NBS1 subnuclear assembly requires the Fanconi anemia C protein. Hum Mol Genet 2002; 11:2531-2546.

45. Wong JC, Buchwald M. Disease model: Fanconi anemia. Trends Mol Med 2002; 8:139-142.

46. Rio P, Segovia JC, Hanenberg H et al. In vitro phenotypic correction of hematopoietic progenitors from Fanconi anemia group A knockout mice. Blood 2002; 100:2032-2039.

47. van de Vrugt HJ, Cheng NC, de Vries Y et al. Cloning and characterization of murine fanconi anemia group A gene: Fanca protein is expressed in lymphoid tissues, testis, and ovary. Mamm Genome 2000; 11:326-331.

48. Wevrick R, Clarke CA, Buchwald M. Cloning and analysis of the murine Fanconi anemia group C cDNA. Hum Mol Genet 1993; 2:655-662.

49. Yang Y, Kuang Y, De Oca RM et al. Targeted disruption of the murine Fanconi anemia gene, Fancg/Xrcc9. Blood 2001; 98:3435-3440.

50. Houghtaling S, Timmers C, Noll M et al. Epithelial cancer in Fanconi anemia complementation group D2 (Fancd2) knockout mice. Genes Dev 2003; 17:2021-2035.

51. Connor F, Bertwistle D, Mee PJ et al. Tumorigenesis and a DNA repair defect in mice with a truncating Brca2 mutation. Nat Genet 1997; 17:423-430.

52. Noll M, Battaile KP, Bateman R et al. Fanconi anemia group A and C double-mutant mice: Functional evidence for a multi-protein Fanconi anemia complex. Exp Hematol 2002; 30:679-688.

53. Chen M, Tomkins DJ, Auerbach W et al. Inactivation of Fac in mice produces inducible chromosomal instability and reduced fertility reminiscent of Fanconi anaemia. Nat Genet 1996; 12:448-451.

54. Whitney MA, Royle G, Low MJ et al. Germ cell defects and hematopoietic hypersensitivity to gamma-interferon in mice with a targeted disruption of the Fanconi anemia C gene. Blood 1996; 88:49-58.

55. Carreau M, Gan OI, Liu L et al. Bone marrow failure in the Fanconi anemia group C mouse model after DNA damage. Blood 1998; 91:2737-2744.

56. Haneline LS, Broxmeyer HE, Cooper S et al. Multiple inhibitory cytokines induce deregulated progenitor growth and apoptosis in hematopoietic cells from Fac-/- mice. Blood 1998; 91:4092-4098.

57. Rathbun RK, Christianson TA, Faulkner GR et al. Interferon-gamma-induced apoptotic responses of Fanconi anemia group C hematopoietic progenitor cells involve caspase 8-dependent activation of caspase 3 family members. Blood 2000; 96:4204-4211.

58. Battaile KP, Bateman RL, Mortimer D et al. In vivo selection of wild-type hematopoietic stem cells in a murine model of Fanconi anemia. Blood 1999; 94:2151-2158.

59. Ludwig T, Chapman DL, Papaioannou VE et al. Targeted mutations of breast cancer susceptibility gene homologs in mice: Lethal phenotypes of Brca1, Brca2, Brca1/Brca2, Brca1/p53, and Brca2/p53 nullizygous embryos. Genes Dev 1997; 11:1226-1241.

60. McAllister KA, Bennett LM, Houle CD et al. Cancer susceptibility of mice with a homozygous deletion in the COOH-terminal domain of the Brca2 gene. Cancer Res 2002; 62:990-994.

61. Scully R, Chen J, Plug A et al. Association of BRCA1 with Rad51 in mitotic and meiotic cells. Cell 1997; 88:265-275.

62. Otsuki T, Furukawa Y, Ikeda K et al. Fanconi anemia protein, FANCA, associates with BRG1, a component of the human SWI/SNF complex. Hum Mol Genet 2001; 10:2651-2660.

63. Bochar DA, Wang L, Beniya H et al. BRCA1 is associated with a human SWI/SNF-related complex: Linking chromatin remodelling to breast cancer. Cell 2000; 102:257-265.

64. Scully R. Interactions between BRCA proteins and DNA structure. Exp Cell Res 2001; 64:67-73.

65. Reuter TY, Medhurst AL, Waisfisz Q et al. Yeast two-hybrid screens imply involvement of Fanconi anemia proteins in transcription regulation, cell signaling, oxidative metabolism, and cellular transport. Exp Cell Res 2003; 289:211-221.

66. Hoatlin ME, Zhi Y, Ball H et al. A novel BTB/POZ transcriptional repressor protein interacts with the Fanconi anemia group C protein and PLZF. Blood 1999; 94:3737-3747.

67. Qiao F, Moss A, Kupfer GM. Fanconi anemia proteins localize to chromatin and the nuclear matrix in a DNA damage and cell cycle-regulated manner. J Biol Chem 2001; 276:23391-23396.

68. Callén E, Ramírez MJ, Creus A et al. The clastogenic response of the 1q12 heterochromatic region to DNA cross-linking agents isindependent of the Fanconi anemia pathway. Carcinogenesis 2002; 23:1267-1271.

69. Surrallés J, Ramírez MJ, Marcos R et al. Clusters of transcription coupled repair in the human genome. Proc Natl Acad Sci USA 2002; 99:10571-10574.

70. Aguilera A. The connection between transcription and genomic instability. EMBO J 2001; 21:195-201.

71. Callen E, Samper E, Ramirez MJ et al. Breaks at telomeres and TRF2-independent end fusions in Fanconi anemia. Hum Mol Genet 2002; 11:439-444.

72. D'Andrea AD, Grompe M. The Fanconi anaemia/BRCA pathway. Nat Rev Cancer 2003; 3:23-34.

73. Godthelp BC, Artwert F, Joenje H et al. Impaired DNA damage-induced nuclear Rad51 foci formation uniquely characterizes Fanconi anemia group D1. Oncogene 2002; 21:5002-5005.

CHAPTER 7

Is Ataxia Telangiectasia a Result of Impaired Coordination between DNA Repair and Cell Cycle Checkpoint Regulators?

Adayabalam S. Balajee and Charles R. Geard

Abstract

Ataxia telangiectasia (AT) is an autosomal recessive multisystem human disorder and patients are characterized by cerebellar ataxia, oculocutaneous telangiectasia, immuno-deficiency, chromosomal instability and radio sensitivity with an increased predisposition to lymphoid cancer in childhood. The gene responsible for AT, ataxia telangiectasia mutated (ATM), has been cloned and its protein product has been biochemically characterized as a serine/threonine kinase belonging to the family of phosphatidylinositol (PI-3) like kinases. Subsequent biochemical studies by several laboratories have identified a number of DNA repair and cell cycle proteins that are phosphorylated by *ATM* kinase in response to different DNA damaging agents. One intriguing question that comes to mind is whether the phenotypic features of AT stem from a DNA repair defect or a cell cycle defect or both. The scope of this review is focused on the potential functions of *ATM* in both DNA repair and cell cycle checkpoint regulation and how deficiencies in these overlapping functions can lead to some of the phenotypic features of AT patients.

Introduction

Many human autosomal recessive disorders display a multisystem impairment reflecting genomic instability manifested at different tissue or organ levels. While the molecular basis for genomic instability remains elusive, it has been hypothesized that a deficiency in the repair of DNA lesions inflicted in the genomic DNA is responsible, at least in certain cases, for genomic instability. Sensitivity of cells derived from certain genomic instability syndromes like xeroderma pigmentosum, Cockayne syndrome, Bloom syndrome, Werner's syndrome and ataxia telangiectasia to different DNA damaging agents seems to favor this hypothesis. Spontaneous and induced chromosomal instability in the form of translocations, sister chromatid exchanges and variegated translocation mosaicism frequently found in cells of these patients indicate deficiencies in DNA repair pathways.

Eukaryotic cells possess diverse DNA repair pathways to deal with a spectrum of DNA lesions induced by endogenous and exogenous agents. Endogenous DNA damage mainly in the form of reactive oxygen species is generated through metabolic activities of cells. Exogenous DNA damage is caused by exposure to ultraviolet light (UV), ionizing radiation

DNA Repair and Human Disease, edited by Adayabalam S. Balajee. ©2006 Landes Bioscience and Springer Science+Business Media.

(IR) and toxic chemicals. While UV induces mainly cyclobutane pyrimidine dimers and pyrimidine-pyrimidone photoproducts, IR induces single strand breaks, double strand breaks, base damage and DNA-protein cross links. UV induced bulky DNA lesions are removed by nucleotide excision repair (NER) pathway while IR induced DNA strand breaks are processed by nonhomologous end joining (NHEJ) and homologous recombination repair (HRR) pathways. Base excision repair (BER) pathway removes the oxidized DNA base lesions generated by oxygen free radicals while mismatch repair and post-replication repair pathways remove mismatched bases arising due to replication errors. In addition to UV induced bulky DNA damage, DNA interstrand cross links generated by anti-tumor drug cisplatin are also processed by NER. Studies on human syndromes that are specifically defective in one or more of these repair pathways have enabled the identification and characterization of proteins involved in various repair pathways.

Cellular sensitivity to DNA damaging agents can result either separately or in combination from a DNA repair deficiency and a failure to block cell cycle progression at G1, S and G2/M phases. Transient cell cycle arrest in response to DNA damage is intended to provide time for the cellular repair machinery to remove the DNA lesions which otherwise may lead to deleterious mutations. Progression of cells through S-phase in the presence of DNA damage can lead to mitotic catastrophe and apoptosis. For example, UV induced apoptosis can occur either before or after the S-phase depending upon the extent of DNA damage as demonstrated for NER deficient Chinese hamster UV61 cells (homologous to human Cockayne syndrome complementation group B (CS-B) cells). UV61 cells progress through G1, S and G2 phases but undergo apoptosis in the subsequent G1 at a low UV dose ($1 J/m^2$) while the cells get permanently arrested at G1 and undergo apoptosis at high UV doses ($>5 J/m^2$). However, repair proficient AA8 cells neither undergo a permanent G1 arrest nor apoptosis at equivalent UV doses. The progression of cells from G1 to S-phase after high UV doses requires an efficient transcription coupled repair, a sub-pathway of NER which is defective in both human CS-B and hamster UV61 cells. Transcription coupled repair is closely linked to basal transcription machinery and this pathway is critical for cellular survival after UV irradiation. A permanent G1 arrest observed in UV 61 cells illustrates that the DNA repair efficiency is critical for the resumption of cell cycle after DNA damage. DNA repair and cell cycle regulation seem to be interdependent on each other and the intrinsic connection between these two processes is beginning to emerge from studies on certain human disorders such as ataxia telangiectasia, whose gene product *ATM* seems to play dynamic roles in the coordinated activities of DNA repair and cell cycle checkpoint regulation in eukaryotic cells. This review is focused on how these two processes are linked by *ATM* kinase and how disruption of this link can lead to genomic instability in AT patients.

Phenotypic Features and Incidence of AT in Human Population

AT is an inherited autosomal recessive human disorder characterized by a multisystem impairment particularly in the nervous system, the immune system and the skin of afflicted individuals. AT occurs with an incidence of 1 per 40,000 live births and the AT patients develop progressive neurological abnormalities in the cerebellum resulting in a staggering gait, mental retardation and severe muscular dysfunction. Dilation of small blood vessels termed as *telangiectasia* is most readily seen in the eye and the skin of AT patients. As many as 10% of the AT patients develop neoplasms before 20 years of age, out of which 88% are lymphoreticluar including Hodgkin's and nonHodgkin's lymphomas. A retrospective study on 234 patients with AT revealed a 61 and 184 fold increase in cancer incidence in whites and blacks respectively with leukemia and lymphoma most prevalent.[1] Heterozygous relatives of individuals with AT are more prone to develop neoplasms than the unrelated individuals. It has been estimated that the allelic loss of the gene responsible for AT accounts for 5 % of all persons who die of cancer prior to 45 years of age. Available evidences indicate that 8.8% of the patients with breast cancer among the white American population are heterozygous for AT. This illustrates that the allelic loss of one AT gene (haploinsufficiency) predisposes the individuals to breast cancer development. Molecular cloning and characterization of the gene responsible for AT phenotype, *ATM* (ataxia telangiectasia mutated) and its protein product related to PI-3 like kinases opened up new exciting possibilities not only to understand the biological functions of this gene but also to explain the basis for the phenotypic features of AT patients.

Cellular Characteristics of AT Patients

One of the hallmarks of AT cells is their increased sensitivity to DNA double strand break inducing agents such as IR and bleomycin. The radiosensitive phenotype was initially suspected when AT patients showed a fatal and severe response when subjected to radiation for the treatment of tumors. Subsequently Taylor et al[2] and many others reported the increased sensitivity of AT cells to IR treatment. AT cells exhibit an elevated level of spontaneous and DNA damage induced chromosome and chromatid type aberrations. Fluorescence in situ hybridization using whole chromosome specific probes for 1, 2 and 4 detected an increased frequency of chromosome translocations in cells derived from both homozygous and heterozygous individuals with AT.[3] Another feature that distinguishes AT cells from normal cells is their inability to block S-phase progression when challenged with IR. AT cells, unlike normal cells, continue to replicate their DNA as judged by the uptake of ^{3}H-Thymidine after IR treatment and this feature is known as radio resistant DNA synthesis (RDS). RDS is a useful marker for assessing the radio sensitivity of cells in vitro. The progression of cells through S-phase in the presence of DNA damage has been considered to be the cause for enhanced chromatid type aberrations involving breaks and gaps observed in AT cells after IR exposure.[4,5] Also, cells from heterozygous individuals with AT show enhanced G2 chromatid aberrations after IR and this elevated sensitivity proves to be a very good diagnostic tool for the identification of heterozygous carriers. In addition to IR, AT cells are also sensitive to topoisomerase inhibitors. Using the topoisomerase I inhibitor, camptothecin (which kills cells in S-phase through the generation of DSBs in replication forks), Johnson et al[6] have shown that AT cells are defective in a sub-pathway of DSB repair that deals with the replication mediated DSBs.

Cell cycle analysis performed on AT cell lines revealed the lack of cell cycle arrest in all cell cycle phases after IR. Lack of G1 arrest in AT cells can be explained by their inability to promptly induce p53 protein[7,8] after IR induced DNA damage. P53 is an important constituent of G1 checkpoint machinery, which transactivates the expression of many downstream targets including a cyclin dependent kinase inhibitor, p21. The requirement for p21 in G1 arrest has been very well established. The defective p53 induction is observed in AT cells only after IR but not after UV irradiation suggesting that the mode of p53 activation is different for different DNA damaging agents.[9,10] As mentioned above, S-phase checkpoint is also defective in AT cells, which exhibit RDS after IR. Although AT cells show a dose-dependent accumulation of G2 cells after IR, G2 arrest is also defective in AT cells and they prematurely enter mitosis with aberrant chromosomes with breaks and gaps. Xu et al[11] have identified two molecularly distinct G2/M phase checkpoints in mammalian cells. The first of these transiently occurs relatively early after IR and represents the failure of G2 cells to progress into M phase and this checkpoint is found to be *ATM* dependent but radiation dose independent. The second one occurs several hours later and represents the cells accumulated in G2 from the preceding cell cycle phases at the time of radiation. The second G2 checkpoint is *ATM* independent but exhibits a dose dependent accumulation of cells in G2 phase. The checkpoint deficiency observed in G1, S and G2 phases makes it difficult to link the radio sensitivity of AT cells with any particular phase of the cell cycle. It appears that the premature entry of G1 cells into

S-phase, S cells into G2 phase and G2 cells into M-phase depending on the cell cycle stage at the time of IR treatment can collectively lead to radiosensitive phenotype of AT cells.

In addition to DNA repair and cell cycle defects, cells derived from AT patients also show abnormal telomere regulation. Accelerated telomere shortening has been demonstrated in the peripheral blood lymphocytes and simian virus 40 transformed fibroblasts of AT patients.[12,13] Furthermore, telomere end fusions were observed in the preleukaemic translocation clones in the T- lymphocytes of AT patients.[12] In accordance with accelerated telomere shortening, *ATM* deficient human and mouse cells in culture undergo premature replicative senescence. Extra-chromosomal telomeric DNA in cells of *ATM* null mice[14] suggests that the *ATM* gene is required for replication and maintenance of chromosomal regions containing telomeric DNA sequences. In yeast, *Saccharomyces cerevisiae*, mutations in Tel1 and Mec1 genes, which are homologous to mammalian *ATM* and ATR (ATM and Rad3 related kinase), result in a high rate of telomeric fusions, dicentrics and circular chromosomes.[15] Additionally, Tel1 and Mec1 double mutants exhibit 80 fold-increased rates of mitotic recombination and chromosome loss. Wong et al[16] have recently examined the functional interaction between telomeres and *ATM* in vivo using the null mice for both *ATM* and telomerase RNA component (Terc). These compound mutants, as expected, showed telomere erosion and genomic instability. However, an increased incidence of T-cell lymphomas, which is often associated with *ATM* deficiency, was eliminated in these compound mutants. Qi et al[17] reported similar observations for *ATM* and telomerase double knock out mice. A proliferation defect detected in all the cells and tissues including the stem/progenitor cells led to the suggestion that *ATM* deficiency and telomere dysfunction act together in destabilizing cellular and whole genome viability. Chronic oxidative stress is considered to be responsible for the accelerated telomere shortening in AT cells and culturing of both normal and AT cells in the presence of anti-oxidant phenyl-butyl-nitrone reduced the rate of telomere shortening.[18] Although telomere dysfunction can explain the premature replicative senescence observed in *ATM* deficient cells, its link with radio sensitivity remains to be validated.

ATM Gene and Its Alterations in Cancer Patients

Savitsky et al[19] cloned the gene responsible for AT condition. *ATM* gene spanning more than 150 kb resides at the chromosomal region 11q23.1. The open reading frame of *ATM* transcript is 9168 nucleotides with a predicted molecular mass of 350 Kd *ATM* protein consisting of 3,056 amino acids.[20] *ATM* protein is related to PI-3 family of kinases (see refs. 20,21, and refs. therein). Uziel et al22 studied the genomic organization of *ATM* gene using long-distance PCR. *ATM* gene is composed of 66 exons, which are spread within a compact region of 150 kb. The first methionine of the open reading frame is located in exon 4 and the stop codon is located in the 3' largest exon of 3.8 kb. The exons range from 43bp to 634 bp, with an average size of 152 bp while the introns vary considerably in size from 100 bp to 11 kb. A recent study on a full length screening of the *ATM* gene in 27 Brazilian families with AT revealed five founder haplotypes that accounted for 55.5% of the families.[23] All mutations (nonsense, splice site and frameshift) led to either truncated or null form of *ATM* protein and these mutations were found in the entire gene without any hotspots.

Several studies have demonstrated a link between *ATM* gene mutations and breast cancer incidence. Within the AT families, breast cancer incidence is increased to 6-7 fold in AT heterozygous carriers suggesting the importance of *ATM* gene in breast cancer development. Broeks et al[24] reported 7 germ line mutations in 82 breast cancer patients who had either early onset of the disease or bilateral breast cancer and concluded that the truncating germ line *ATM* mutations contribute to an early onset of breast cancer. However, FitzGerald et al[25] who detected *ATM* mutations in only 2 of 401 breast cancer patients of approximately 40 years of age concluded that the truncating germ line *ATM* mutations do not contribute to an early breast cancer development. To clarify these observations and to define the whole spectrum of *ATM* gene mutations, Dork et al[26] screened 1000 breast cancer patients and 500 randomly chosen individuals for *ATM* gene mutations. Their study revealed 21 distinct sequence alterations throughout the coding region and 1 common splicing mutation in intron 10. A total of 46% patients showed 1 of 16 different amino acid substitutions and 1.6% of the patients showed a truncating mutation. A recent study identified a ser49cys variant of the *ATM* gene to be a common feature in patients with breast carcinoma.[27] Two *ATM* mutations [7271T→G and IVS10-6T→G] have been suggested to confer cumulative breast cancer risks of 55 and 78% by 70 years of age in the Australian families with breast cancer[28] although Szabo et al[29] found that these mutations do not contribute significantly to elevated breast cancer incidence.

Thorstenson et al[30] identified 137 *ATM* gene alterations in 270 hereditary breast and ovarian cancer families that were previously screened for mutations in the breast cancer susceptibility genes, BRCA1 and BRCA2. A total of 7 (5 leading to truncated *ATM* protein and 2 missense mutations in the catalytic kinase domain of the highly conserved carboxy terminus of the protein) out of 137 mutations were presumed to result in ataxia telangiectasia depending on their effect on *ATM* protein. These 7 mutations were found in 10 families. This study suggests a higher prevalence of *ATM* mutations in breast and ovarian cancer families. Loss of heterozygosity at the region of *ATM* gene location (11q22-23) has been frequently associated with sporadic lymphoid tumors and a high prevalence of the *ATM* gene alterations have been reported in sporadic lymphoproliferative disorders.[31] These findings indicate the contribution of *ATM* gene to the pathogenesis of these tumors. Aberrant methylation of the *ATM* promoter has also been demonstrated in 25 % of the head and neck squamous cell carcinomas[32] indicating the reduced *ATM* protein expression may be causally linked with this tumor type. Alterations in the *ATM* gene observed in breast, ovarian, and head and neck squamous cell carcinomas seem to suggest that it may act as a tumor suppressor for certain tumor types. Further studies are required to clarify whether or not *ATM* gene alteration is a primary cause for cancer development.

ATM Protein

Diverse phenotypic features in AT patients and a high occurrence of *ATM* gene mutations in a wide variety of cancer patients illustrate that its gene product must play vital roles in the maintenance of genomic stability. As mentioned earlier, *ATM* gene encodes a protein product of 360kd that belongs to a family of PI-3 kinases. Other members of *ATM* related group of PI-3 kinases include Tel1 and Mec1 of Saccharomyces cerevisiae, Mec1 homologs of Rad3 Schizosaccharomyces pombe (Rad3), Drosophila melanogaster (Mei-41) and humans (ATR). Cells deficient in these proteins, like AT cells, display hypersensitivity to DNA damaging agents and lack of cell cycle checkpoint regulation upon DNA damage. In addition to kinase domain, *ATM* protein contains a proline rich region and a leucine zipper region both of which implicate a role for *ATM* in signal transduction pathways. *ATM* is a phosphoprotein and is detected in all the tissues with a higher level of expression in testis, spleen and thymus.[33] *ATM* protein is predominantly nuclear in distribution[34,35] but in certain cell types a significant fraction is also found in cytoplasmic vesicles.[36,37] Ectopic expression of full-length *ATM* cDNA in *ATM* deficient cells rescues them from radio sensitivity and complements the cell cycle checkpoint defects in response to IR.[34,38] Restoration of cell survival, S-phase arrest and reduced chromosomal aberrations after IR in AT cells transfected only with carboxy terminal fragment of *ATM* that contains the PI-3 kinase related domain demonstrates that the majority of *ATM* functions is intrinsically associated with its kinase domain.[39] Atomic force microscopy reveals the existence of two populations of *ATM* protein as monomeric and tetrameric states with a preferential binding to DNA ends.[40] Studies over the years have identified a number of proteins involved in DNA repair and cell cycle that are phosphorylated by *ATM* kinase. Identification of potential targets for *ATM* kinase greatly enables us to gain insights into not only how *ATM* functions in the cellular environment but also the biological consequences of its absence.

Role of ATM Kinase in Cell Cycle Checkpoint Regulation

One of the well-known responses to IR treatment is the activation of a nuclear tyrosine kinase encoded by c-abl proto-oncogene. Activation of c-Abl tyrosine kinase was defective in cells derived from *ATM* deficient mice and the ectopic expression of *ATM* kinase domain restored the activation. Further, *ATM* kinase has been shown to phosphorylate tyrosine kinase in vitro on serine 465.[41] *ATM* protein exhibits DNA stimulated kinase activity that phosphorylates serine 15 at the amino terminal region of p53[40,42] and IR treatment of cells enhances the p53 directed ser15 phosphorylation by *ATM* kinase.[43] P53 is a tumor suppressor protein that plays an important role in genome surveillance through transcriptional activation of a number of downstream targets such as p21, Gadd45 and Bax. Ectopic expression of *ATM* protein in AT cells restores the IR induced p53 phosphorylation while anti-sense *ATM* expression in normal cells abolishes the p53 phosphorylation demonstrating the requirement of *ATM* for IR induced p53 phosphorylation.[44] In an effort to determine the region(s) of interaction between p53 and ATM, Khanna et al[44] analyzed the ability of different regions of GST-ATM fusion protein to bind to[35] S-labelled, in vitro translated full length p53 protein and found that p53 bound strongly to GST-ATM construct containing the residues 1-246 and weakly to amino acid residues 2862-3012 of the PI-3 kinase domain. In addition to ser15, *ATM* phosphorylates serines 46 and 9 of p53 in response to IR. Phosphorylation of ser15 together with serines 6,9 and threonine 18 are all required to stabilize and activate sequence specific DNA binding of p53 via enhancing the C-terminal acetylation at lysine 320 and 382 residues.[45] Mdm2 is a negative regulator of p53 and *ATM* phosphorylates Mdm2 on serine 395 in response to IR and the phosphorylation of Mdm2 occurs earlier than p53 activation.[46] *ATM* kinase thus acts at two levels to ensure stability and activation of p53 in response to DNA damage: at the first level, *ATM* inactivates the inhibitory effect of Mdm2 on p53 through phosphorylation and at the second level it enhances the p53 stability by phosphorylating its serine residues. A fine orchestration of these two events by *ATM* kinase enables the cells to undergo either G1 checkpoint control or apoptosis depending on the extent of DNA damage.

ATM kinase also specifically phosphorylates BRCA1 in a region rich in serine- glutamine residues in response to DSBs induced by IR.[47] BRCA phosphorylation is functionally relevant as the BRCA1 protein mutated in the phosphorylation sites failed to rescue the hypersensitivity of BRCA1 deficient cells. Gatei et al,[48] using phospho-specific antibodies demonstrated that serine residues 1387, 1423 and 1457 of BRCA1 are phosphorylated by *ATM* kinase. Interestingly, mutation of serine 1423 in BRCA1 abrogates G2/M checkpoint but not the IR induced S-phase checkpoint while mutation in serine 1387 abolishes the S-phase checkpoint but not the G2/M checkpoint.[49] Thus, two post-translational modifications of BRCA1 protein induced by *ATM* kinase enforce S- and G2/M specific checkpoint controls in mammalian cells. *ATM* dependent phosphorylation of BRCA1 may be critical for dealing with DSBs that spontaneously arise in cells during replication and recombination and the failure of their adequate repair may lead to genomic instability and cancer. Increased incidence of breast cancer in AT families seems to support such an assumption. *ATM* kinase phosphorylates BRCA1 associated protein, CtIP at serine residues 664 and 745 and the hyperphosphorylated CtIP dissociates from BRCA1 upon IR treatment. Mutation of these serine residues to alanine abrogates the CtIP dissociation from BRCA1 and represses BRCA1 mediated induction of Gadd45.[50] It is thus apparent that the DNA damage response to DSBs is lost in the cells in the absence of either *ATM* or BRCA1 and as a consequence, cells deficient in either are hypersensitive to IR.

Nijmegen breakage syndrome (NBS) is a hereditary disorder and the patients with NBS share features with AT such as immunodeficiency, cancer predisposition and radio sensitivity. The gene responsible for this disorder, Nbs1 has been cloned and Nbs1deficient cells also display RDS in response to IR without any cell cycle checkpoint defects like that observed in AT cells. The link between NBS and AT became evident by the finding of Nbs1 phosphorylation on serine 143 by *ATM* kinase upon IR treatment.[51] Although Nbs1 gene product is not required for *ATM* kinase activation and *ATM* dependent DNA damage responses, mutation introduced in Nbs1 protein at the site of *ATM* phosphorylation abrogates the S-phase checkpoint in cells. Nbs1 is a component of MRN (Mre11-Rad50-Nbs1) complex involved in the recognition/repair of DNA DSBs as well as *ATM* mediated cell cycle checkpoints in response to DNA damage. Although IR activates *ATM* kinase, the precise activation mechanism remained obscure until Bakkenist and Kastan[52] identified an autophosphorylation site on serine 1981 of the *ATM* protein. IR triggers a rapid *ATM* autophosphorylation resulting in the dissociation of dimers and initiation of *ATM* kinase activity. The authors have suggested that the *ATM* activation is not dependent on direct binding to DSBs but may result from changes in the chromatin structure. The autophosphorylation activity of *ATM* seems to be dependent on the functional status of Nbs1 gene product as the kinetics and the extent of serine 1981 *ATM* phosphorylation are altered in Nbs1 deficient cells.[53] Transfection of Nbs1 cells with wild type Nbs1 gene as well as mutants lacking FHA domain and *ATM* phosphorylation site rescued the timely phosphorylation of not only *ATM* serine 1981 but also another target of ATM, Smc1 (structural maintenance of chromosome 1 protein). Interestingly Nbs1 lacking the Mre11 interacting domain is unable to rescue the autophosphorylation of *ATM* indicating the importance of MRN complex in *ATM* kinase activation. Earlier studies by Kim et al[54] and Yazdi et al[55] demonstrated that *ATM* phosphorylates Smc1 protein on serines 957 and 966 in vitro and in vivo and mutations in these sites abolish IR induced S-phase checkpoint. Similar to Chk1 and *ATM* autophosphorylation, optimal phosphorylation of Smc1 by *ATM* kinase requires a functional Nbs1 gene product. However, Smc1 phosphorylation at the same sites triggered by UV and hydroxyurea occurs independent of *ATM* gene product implicating the involvement of a kinase other than ATM.

Yet another important target for *ATM* kinase is Chk2 which is the mammalian homolog of budding yeast Rad53 and fission yeast Cds1. Chk2 is phosphorylated at threonine 68 by *ATM* kinase in response to IR but not to UV and hydroxyurea.[56,57] Yeast Cds1 counterpart of human, designated as HuCds1, requires a functional *ATM* gene product for phosphorylation after IR but not hydroxyurea.[58] IR induced phosphorylation of Chk1 at serine 317 is dependent on *ATM* but also requires a functional Nbs1 gene product.[59] IR induced phosphorylation of *ATM* dependent targets, Chk2, RPA p34 subunit and p53 ser15 was also reduced in Nbs1 deficient cells despite a normal activation of *ATM* kinase in response to IR.[60] This finding suggests an accessory role for Nbs1 in promoting an efficient *ATM* dependent phosphorylation of proteins involved in G1/S arrest in response to IR induced DNA damage. Human Rad9, a homolog of fission yeast Schizosaccharomyces pombe, is an important cell cycle checkpoint regulator and *ATM* phosphorylates human Rad9 on serine 272 in response to IR.[61] Expression of Rad9 mutant with an alanine substitution for serine 272 in human lung fibroblasts sensitizes the cells to radiation and abolishes the IR induced G1/S checkpoint. Rad9 of budding yeast, which is different from that of fission yeast, requires a functional Tel1 and Mec1 (homologs of human *ATM* and ATR) gene products for IR induced hyperphosphorylation.[62] It is interesting to note that in asynchronous cultures, Rad9 hyperphosphorylation requires Tel1 and Mec1 while Rad17, Rad24, Mec3 and Ddc1 proteins are needed for damage dependent phosphorylation in G1 arrested cells. This clearly illustrates a cell cycle specific participation of different checkpoint genes in response to DNA damage.

The link between *ATM* deficiency and abnormal telomere dynamics is supported by the finding of phosphorylation of telomeric protein Pin2/TRF1 by *ATM* kinase on serine129-Gln site in response to IR.[63] Pin2/TRF1 is a negative regulator of telomere elongation and is expressed in a cell cycle specific manner. Expression of Pin2 mutant refractory to *ATM* kinase phosphorylation enhances the radio sensitivity of AT cells by speeding up the entry into mitosis and apoptosis. Thus, the *ATM* kinase regulates the biological activity of Pin2/TFR1 through phosphorylation and a failure to modulate its activity in *ATM* deficient cells may be responsible for radiation hypersensitivity and telomere loss.

Role of ATM Kinase in DNA Repair

In addition to proteins involved in cell cycle checkpoint regulation, *ATM* kinase also phosphorylates important proteins that participate in double strand break repair. A histone H2A variant, H2AX is phosphorylated on serine 139 in response to IR and the phosphorylated γ-H2AX binds to the DSB sites.[64] γ-H2AX is considered to facilitate double strand break repair by recruiting the repair factors to the lesion sites. γ-H2AX form foci in irradiated cells and the γ-H2AX foci colocalize with BRCA1, Rad50 and Rad51 proteins.[65] Cells derived from H2AX null mice are radiosensitive with deficiencies in IR induced foci formation of BRCA1, Nbs1 and p53 binding protein1.[66,67] *ATM* Kinase is responsible for H2AX phosphorylation and *ATM* deficient cells show greatly attenuated induction of γ-H2AX after IR. A recent study has shown the colocalization of *ATM* and γ-H2AX at the sites of DSBs in human cells indicating the concerted action of these two proteins in DSB repair.[68] Replication stress also triggers H2AX phosphorylation on serine 139 but by the ATR kinase rather than ATM. This illustrates how specific the activation is for different kinases in response to different DNA damaging signals.

Replication protein A (RPA), a single strand binding protein, plays important roles in diverse DNA metabolic activities such as DNA replication, repair and recombination. RPA participates in the DNA damage recognition step of nucleotide excision repair and is also implicated in the long patch base excision repair pathway. RPA occurs in a heterotrimeric state consisting of three sub-units with molecular weights of 70kd, 34kd and 14kd. Initial indication for a probable functional interaction between RPA and *ATM* has emerged from the study of Plug et al[69] who showed that RPA and *ATM* proteins colocalize along the length of synapsed meiotic chromosomes and at sites where interactions between ectopic homologous chromosome regions tend to initiate. Among the three subunits, RPAp34 is a phosphoprotein, which is phosphorylated in response to DNA damage and replication blockage. In budding yeast, RPA is hyperphosphorylated by the *ATM* homolog, Mec1[70] and RPA phosphorylation is also attenuated in *ATM* deficient human cells. RPA phosphorylation thus represents a conserved DNA damage response pathway in eukaryotes, which is mediated by *ATM* kinase. In addition to *ATM* kinase, involvement of DNA-PK in RPA phosphorylation initiated by camptothecin induced replication blockage has also been shown.[71] Hence a coordinated action of *ATM* and DNA-PK is required for RPA phosphorylation depending upon the nature of signal.[72] Despite the fact that AT cells are not overly sensitive to UV radiation, UV induced RPA phosphorylation is dependent on the expression of *ATM* protein.[73]

53BP1 is a human BRCT containing protein and it was originally identified as a p53-interacting protein.53BP1 forms foci in response to IR and 53BP1 foci colocalize with Mre11 and phosphorylated form of H2AX. 53BP1 is hyperphosphorylated in response to X-rays and inhibitors of ATM-kinase reduced the phosphorylation of 53BP1 indicating the involvement of *ATM* kinase in its DNA damage dependent phosphorylation. Also, cells deficient in *ATM* display reduced phosphorylation of 53BP1 in response to IR treatment.[74]

53BP1 is also implicated in G2/M phase arrest imposed by IR and G2/M arrest is achieved through binding of phosphorylated 53BP1 with γ-H2AX.

Recent studies have suggested that MRN complex may amplify the kinase activity of ATM.[53,75] These studies have shown the requirement of MRN complex for *ATM* kinase activation in vitro. Phosphorylation of some of the *ATM* targets such as p53, Chk2 and H2AX is enhanced by MRN complex and the phosphorylated form of Nbs1 is critical for the stimulation of Chk2 but not p53 phosphorylation by *ATM* kinase.[75] Although not absolutely required for *ATM* kinase activation, MRN complex may help *ATM* kinase to attain its peak activity so that it can effectively participate in genome surveillance pathway by phosphorylating a number of downstream targets. MRN complex appears to be critical for DSB repair as cells mutated in any of the proteins in the complex exhibit a radiosensitive phenotype. Hence, MRN dependent activation of *ATM* kinase attests a role for *ATM* in DNA damage response pathway.

Two major pathways in mammalian cells repair DSBs generated by IR treatment: (i) Nonhomologous end joining (NHEJ) pathway and (ii) homologous recombination repair (HRR) pathway. However, the relative contribution of these two pathways is not clearly worked out. In yeast, HRR seems to be predominant, while in higher eukaryotes both of these participate in the removal of DSBs. MRN complex is considered to be an important component of HRR and the molecular link between MRN complex and *ATM* kinase therefore suggests a role for *ATM* in HRR pathway. In support, Golding et al[76] have demonstrated a requirement for *ATM* in the regulation of HRR not only in S-phase but also in G2 phase of the cell cycle. As stated earlier, *ATM* deficient cells are defective in a sub-pathway of repair that removes the DSBs that are generated by camptothecin in newly replicating DNA. It is not known whether or not HRR is involved in removing DSBs induced in the replicating forks.

Ataxia Telangiectasia: Cell Cycle versus DNA Repair Defects

It is evident from the foregoing account that *ATM* kinase participates both in DNA repair and cell cycle checkpoint control through phosphorylation of a number of downstream targets. The question of whether the radiosensitivity of AT cells stems primarily from a repair deficiency or a failure to halt cell cycle in response to DNA damage remains largely unanswered. It is not clear whether these two processes occur simultaneously or one precedes the other. It would be interesting to determine whether cells deficient in cell cycle regulation are more radiosensitive than DNA repair deficient cells. Furthermore, it is equally important to know which phase of cell cycle checkpoint defect (G1, S and G2/M) confers radiosensitivity. Human cells immortalized by either simian virus 40 or Epstein Barr virus, which lack G1 checkpoint regulation owing to inactivation of p53 dependent pathway, are not overly sensitive to radiation at permissible dose ranges as compared to primary fibroblasts with an intact G1 checkpoint control. Mice deficient in Chk2 kinase, which is thought to mediate *ATM* dependent DNA damage signaling, are radio resistant in spite of a defective G1/S checkpoint control.[77] In contrast, cells with mutations in S-phase checkpoint genes are universally sensitive to all qualities of radiation. These include cells derived from patients inflicted with AT, NBS and AT-like Disorder (ATLD). Also cells deficient in genes that regulate S-phase in response to DNA damage such as BRCA1, Chk2 and ATR are radiosensitive. In addition to S-phase defect, *ATM* cells are unable to progress from G2 to M phase after low doses of IR. It is becoming increasingly clear that disruption of either *ATM* gene or one of its kinase substrates predisposes the cells to radiosensitivity.

IR induced double strand breaks are rejoined by both fast and slow components of repair activity. The fast component has a t 1/2 of 30 min while the slow component has a t

1/2 of 2-4 hours. Since the majority of DSBs is repaired by the fast component, it is logical to presume that the cell cycle checkpoint control is probably required for the slow repair component. The factors contributing to the fast repair component are not clear but this fast repair activity most likely depends on chromatin structure and DNA sequence context. Future research should focus on identifying the genomic locations of the DSBs that are subjected to fast and slow components of repair. Cells defective in NHEJ show reduced DSB repair throughout the cell cycle phases while in HR defective cells, DSB repair is greatly impaired in S-phase and to a lesser extent in G2 phase.[78] Rejoining of DSBs appears to be normal in AT cells although the fidelity of rejoining is probably low because chromosome and chromatid type aberration frequencies were higher in AT cells than normal cells. Irradiation of plateau phase AT cells at 4°C with 30 Gy of gamma rays resulted in a faster repair of DSBs during the first 6hr period yet the residual damage at 24hrs after IR was higher in AT cells.[79] As these experiments were carried out in plateau phase cells, lack of cell cycle checkpoint control may not be responsible for the increased residual damage in AT cells. The authors have contended that the defective repair of DSB is a mechanism for radiosensitivity in AT. A recent study employing pulse field gel electrophoresis and γ-H2AX foci labeling showed the persistence of a subset of DSBs in plateau phase AT cells[80] which was more than that observed in cells defective in DNA ligase IV. As stated before, Johnson et al[6] demonstrated a deficiency in a sub pathway of DSB repair that deals with replication mediated DSBs generated by camptothecin. Thus, it appears that AT cells are deficient in processing a subset of DSBs irrespective of the cell cycle phases. It is intriguing as to why AT cells are unable to repair a subset of DSBs and how these unrepaired DSBs confer radiosensitivity. A likely explanation is that these are clustered DSBs with strand breaks and complex oxidized base lesions in close proximity which are difficult to repair by AT cells owing to a deficiency in base excision repair pathway. Characterizing the nature of these lesions and their genomic location are of critical importance in understanding the DNA repair basis of radiosensitivity in AT cells. Although different potential protein factors are identified in IR induced DNA damage signaling pathway, the mechanism by which the DNA damage is recognized, signaled and transduced is not resolved. A direct role for *ATM* in DSB repair is debatable despite the observation of its direct binding to a well-known DSB binding factor, γ-H2AX. However, γ-H2AX is shown to be dispensable for the initial recognition of DSBs despite its importance for the foci formation of Nbs1, BRCA1 and 53BP1 after IR.

Radio sensitization of mammalian cells by different kinase inhibitors provides support for the involvement of cell cycle checkpoint control in DSB repair activity. For example, caffeine, which is an inhibitor of ATM/ATR kinases, sensitizes cells to the radiation effects by abrogation of G2/M phase checkpoint. However, caffeine-induced radio sensitization can also occur independent of the lack of G2/M phase arrest in human bladder cancer cells.[81] Furthermore, homologous recombination repair pathway rather than NHEJ seems to be the target for caffeine induced radio sensitization because cells deficient in HR pathway could not be efficiently sensitized by caffeine.[82,83] Similarly, wortmannin, an inhibitor of PI-3 like kinases also enhances radiosensitivity of mammalian cells. Although abrogation of cell cycle checkpoints leads to radiosensitivity, a defect in DNA repair may be primarily responsible for radiosensitivity in AT cells. This notion is greatly strengthened by the repair deficiency detected in noncycling AT cells where the loss of checkpoint control is not likely to have any consequences. Hence the radiosensitivity in noncycling AT cells may be due to a repair deficiency while the radiosensitivity of cells in S and G2 phases may be due to DNA repair deficiency superimposed on cell cycle checkpoint control deficiencies.

A schematic diagram illustrating the role of ATM/ATR kinases in the regulation of different cell cycle phases in response to IR is given in Figure 1.

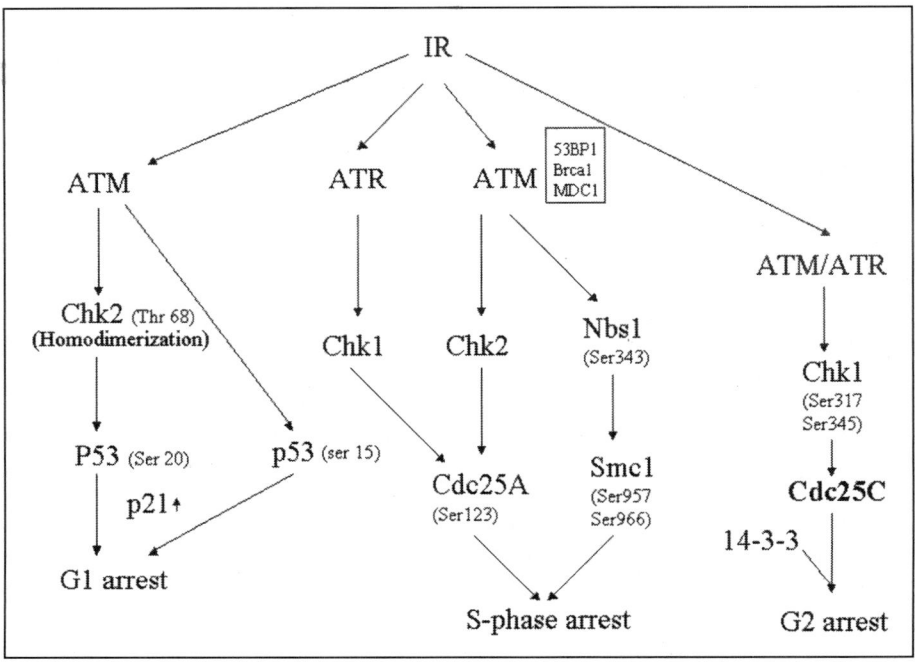

Figure 1. Regulation of cell cycle checkpoint control by ATM/ATR kinases.

ATM kinase activated by IR phosphorylates Chk2, which in turn phosphorylates serine 20 of p53. G1 arrest is subsequently achieved by transactivation of p53 dependent genes including a CDK inhibitor, p21. Alternatively, *ATM* can directly phosphorylate p53 on serine 15 and activate p53 dependent G1/S arrest. Intra S-phase checkpoint is achieved by a combination of *ATM* and ATR kinases. Two *ATM* dependent pathways are operative for the intra-S-phase checkpoint control. In the first pathway, *ATM* can phosphorylate either Chk1 or Chk2, which in turn results in the intra-S-phase checkpoint through phosphorylation of CDC25A.In the second pathway, S-phase checkpoint is achieved by phosphorylation of Nbs1 on serine 343 and SMC1 on serines 957 and 966 by ATM. Phosphorylation of SMC1 by *ATM* however requires a functional Nbs1 gene product as Nbs1 cells failed to show SMC1 phosphorylation after IR. A complex of 53BP1, BRCA1 and MDC1, which is shown to modulate the phosphorylation of Chk1, Chk2 and Nbs1, is also required for the intra-S-phase checkpoint. *ATM* and ATR kinases activate Chk1 by phosphorylation on serine 317 and serine 345 and the activated Chk1 in turn phosphorylates CDC25C on serine 217. Phosphorylation of CDC25C creates a binding site for 14-3-3 and 14-3-3 binding prevents the dephosphorylation of CDC25C as well as the activation of mitotic kinase complex cyclin B-Cdc2, resulting in the blockage of cells at G2.

Increased chromosome and chromatid breaks observed in AT cells after IR suggests that the loss of cell cycle checkpoint is a primary cause for chromosomal instability. Failure to repair a subset of DSBs in the genomic DNA coupled with the premature progression of damaged S-and G2-phase cells into mitosis is a likely mechanism for increased chromosomal aberrations in AT cells.

Genotype and Phenotype Correlation of AT Patients

It is intriguing as to how a single mutated gene can result in a wide variety of phenotypic traits in patients afflicted with ataxia telangiectasia. As mentioned before, AT patients display cerebellar ataxia, oculocutaneous telangiectasia, immunodeficiency, chromosomal instability and radio sensitivity with an increased predisposition to lymphoid cancer in childhood. A combination of DSB repair deficiency and loss of cell cycle checkpoint control can easily explain the radiosensitivity as well as spontaneous and IR induced chromosomal instability of AT patients. Likewise, chromosomal translocations arising due to lack of repair of a subset of DSBs may contribute to an increased frequency of lymphoid cancer in childhood. One of the characteristic features of AT patients is the neurological abnormality involving the degeneration of Purkinje cells and granule neurons in the cerebellar cortex. The brain cells are post mitotic and hence the neurodegeneration cannot solely be explained by abnormal cell cycle checkpoints in response to DNA damage. The hypothesis that the oxidative DNA damage accumulation in the metabolically active brain cells is a causative factor for neurodegeneration has received a great deal of attention in recent years. Consistent with this possibility, a few recent studies have indicated that *ATM* may act as a sensor of oxidative damage and stress. Increased levels of oxidized macromolecules (proteins and lipids) and the heme oxygenase have been reported in cerebellum, cerebral cortex and testes of the *Atm*[−/−] mice.[84] The localization of *ATM* protein in cytoplasmic peroxisomes, which contain enzymes involved in a number of metabolic processes including the peroxide based respiration, also suggests a role for *ATM* in maintaining the steady state level of oxidative influx in the cellular environment.[36] Several cell lines of AT showed decreased level of catalase and increased levels of lipid hydroperoxides.[36] AT cells exhibit increased sensitivity to undergo apoptotic death after treatment with oxidative DNA damaging agents.[85] Although these studies point to defects in oxidative damage processing in AT, a direct role for *ATM* in repair of oxidative damage is yet to be established. Future research should focus on whether or not AT cells are deficient in DNA repair pathways other than DSB.

Conclusions

ATM is being increasingly realized as an important kinase as many of its downstream substrates play vital roles in a wide array of DNA metabolic activities like replication, transcription, repair and recombination. Hence disruption of phosphorylation of *ATM* targets by *ATM* deficiency can differentially affect various cellular processes leading to a multisystem impairment in AT patients. Substantial progress has been made in recent years in the identification and characterization of a number of targets for *ATM* kinase, which greatly improved our understanding of the regulatory functions of *ATM* in DSB repair and cell cycle checkpoint control. However, elucidating the exact role of *ATM* in radiosensitivity and neurodegeneration should be a major focus of future research, which will help in the design of therapeutic drugs to minimize the lethal effects of radiosensitivity and neurodegeneration in AT patients.

Acknowledgements

I apologize to those whose research papers could not be cited due to space limitation. The financial support awarded to Dr. Charles R. Geard from DHHS, NIH (CA 75061, CA 49062 and RR-11623) and DOE (DEFG0298ER62687) Office of Science (BER) is gratefully acknowledged. This article is dedicated to Prof. Tikaram Sharma, Banaras Hindu University, Varanasi, India on the occassion of his 70th birthday.

References

1. Morrell D, Cromartie E, Swift M. Mortality and cancer incidence in 263 patients with ataxia-telangiectasia. J Natl Cancer Inst 1986; 77:89-92.
2. Taylor AM, Harnden DG, Arlett CF et al. Ataxia telangiectasia: A human mutation with abnormal radiation sensitivity. Nature 1975; 258:427-9.
3. Stumm M, Neubauer S, Keindorff S et al. High frequency of spontaneous translocations revealed by FISH in cells from patients with the cancer-prone syndromes ataxia telangiectasia and Nijmegen breakage syndrome. Cytogenet Cell Genet 2001; 92:186-91.
4. Beamish H, Lavin MF. Radiosensitivity in ataxia-telangiectasia: Anomalies in radiation-induced cell cycle delay. Int J Radiat Biol 1994; 65:175-84.
5. Paules RS, Levedakou EN, Wilson SJ et al. Defective G2 checkpoint function in cells from individuals with familial cancer syndromes. Cancer Res 1995; 55:1763-73.
6. Johnson RT, Gotoh E, Mullinger AM et al. Targeting double-strand breaks to replicating DNA identifies a subpathway of DSB repair that is defective in ataxia-telangiectasia cells. Biochem Biophys Res Commun 1999; 261:317-25.
7. Kastan MB, Zhan Q, el-Deiry WS et al. A mammalian cell cycle checkpoint pathway utilizing p53 and GADD45 is defective in ataxia-telangiectasia. Cell 1992; 71:587-97.
8. Morgan SE, Kastan MB. p53 and ATM: Cell cycle, cell death, and cancer. Adv Cancer Res 1997; 71:1-25.
9. Khanna KK, Lavin MF. Lonizing radiation and UV induction of p53 protein by different pathways in ataxia-telangiectasia cells. Oncogene 1993; 8:3307-12.
10. Canman CE, Wolff AC, Chen CY et al. The p53-dependent G1 cell cycle checkpoint pathway and ataxia-telangiectasia. Cancer Res 1994; 54:5054-8.
11. Xu B, Kim ST, Lim DS et al. Two molecularly distinct G(2)/M checkpoints are induced by ionizing irradiation. Mol Cell Biol 2002; 22:1049-59.
12. Metcalfe JA, Parkhill J, Campbell L et al. Accelerated telomere shortening in ataxia telangiectasia. Nat Genet 1996; 13:350-3.
13. Xia SJ, Shammas MA, Shmookler Reis RJ et al. Reduced telomere length in ataxia-telangiectasia fibroblasts. Accelerated telomere shortening in ataxia telangiectasia. Mutat Res 1996; 364:1-11.
14. Hande MP, Balajee AS, Tchirkov A et al. Extra-chromosomal telomeric DNA in cells from Atm(-/-) mice and patients with ataxia-telangiectasia. Hum Mol Genet 2001; 10:519-28.
15. Craven RJ, Greenwell PW, Dominska M et al. Regulation of genome stability by TEL1 and MEC1, yeast homologs of the mammalian ATM and ATR genes. Genetics 2002; 161:493-507.
16. Wong KK, Maser RS, Bachoo RM et al. Telomere dysfunction and ATM deficiency compromises organ homeostasis and accelerates ageing. Nature 2003; 421:643-8.
17. Qi L, Strong MA, Karim BO et al. Short telomeres and ataxia-telangiectasia mutated deficiency cooperatively increase telomere dysfunction and suppress tumorigenesis. Cancer Res 2003; 63:8188-96.
18. Tchirkov A, Lansdorp PM. Role of oxidative stress in telomere shortening in cultured fibroblasts from normal individuals and patients with ataxia-telangiectasia. Hum Mol Genet 2003; 12:227-32.
19. Savitsky K, Bar-Shira A, Gilad S et al. A single ataxia telangiectasia gene with a product similar to PI-3 kinase. Science 1995; 268:1749-53.
20. Savitsky K, Sfez S, Tagle DA et al. The complete sequence of the coding region of the ATM gene reveals similarity to cell cycle regulators in different species. Hum Mol Genet 1995; 4:2025-32.
21. Zakian VA. ATM-related genes: What do they tell us about functions of the human gene? Cell 1995; 82:685-7.
22. Uziel T, Savitsky K, Platzer M et al. Genomic organization of the ATM gene. Genomics 1996; 33:317-20.
23. Coutinho G, Mitui M, Campbell C et al. Five haplotypes account for fifty-five percent of ATM mutations in Brazilian patients with ataxia telangiectasia: Seven new mutations. Am J Med Genet 2004; 126A:33-40.
24. Broeks A, Urbanus JH, Floore AN et al. ATM-heterozygous germline mutations contribute to breast cancer-susceptibility. Am J Hum Genet 2000; 66:494-500.
25. FitzGerald MG, Bean JM, Hegde SR et al. Heterozygous ATM mutations do not contribute to early onset of breast cancer. Nat Genet 1997; 15:307-10.

26. Dork T, Bendix R, Bremer M et al. Spectrum of ATM gene mutations in a hospital-based series of unselected breast cancer patients. Cancer Res 2001; 61:7608-15.
27. Buchholz TA, Weil MM, Ashorn CL et al. A Ser49Cys variant in the ataxia telangiectasia, mutated, gene that is more common in patients with breast carcinoma compared with population controls. Cancer 2004; 100:1345-51.
28. Chenevix-Trench G, Spurdle AB, Gatei M et al. Dominant negative ATM mutations in breast cancer families. J Natl Cancer Inst 2002; 94:205-15.
29. Szabo CI, Schutte M, Broeks A et al. Are ATM mutations 7271T—G and IVS10-6T—G really high-risk breast cancer-susceptibility alleles? Cancer Res 2004; 64:840-3.
30. Thorstenson YR, Roxas A, Kroiss R et al. Contributions of ATM mutations to familial breast and ovarian cancer. Cancer Res 2003; 63:3325-33.
31. Gumy-Pause F, Wacker P, Sappino AP. ATM gene and lymphoid malignancies. Leukemia 2004; 18:238-42.
32. Ai L, Vo QN, Zuo C et al. Ataxia-telangiectasia-mutated (ATM) gene in head and neck squamous cell carcinoma: Promoter hypermethylation with clinical correlation in 100 cases. Cancer Epidemiol Biomarkers Prev 2004; 13:150-6.
33. Chen G, Lee E. The product of the ATM gene is a 370-kDa nuclear phosphoprotein. J Biol Chem 1996; 271:33693-7.
34. Ziv Y, Bar-Shira A, Pecker I et al. Recombinant ATM protein complements the cellular A-T phenotype. Oncogene 1997; 15:159-67.
35. Watters D, Khanna KK, Beamish H et al. Cellular localisation of the ataxia-telangiectasia (ATM) gene product and discrimination between mutated and normal forms. Oncogene 1997; 14:1911-21.
36. Watters D, Kedar P, Spring K et al. Localization of a portion of extranuclear ATM to peroxisomes. J Biol Chem 1999; 274:34277-82.
37. Brown KD, Ziv Y, Sadanandan SN et al. The ataxia-telangiectasia gene product, a constitutively expressed nuclear protein that is not up-regulated following genome damage. Proc Natl Acad Sci USA 1997; 94:1840-5.
38. Zhang N, Chen P, Khanna KK et al. Isolation of full-length ATM cDNA and correction of the ataxia-telangiectasia cellular phenotype. Proc Natl Acad Sci USA 1997; 94:8021-6.
39. Morgan SE, Lovly C, Pandita TK et al. Fragments of ATM which have dominant-negative or complementing activity. Mol Cell Biol 1997; 17:2020-9.
40. Smith GC, Cary RB, Lakin ND et al. Purification and DNA binding properties of the ataxia-telangiectasia gene product ATM. Proc Natl Acad Sci USA 1999; 96:11134-9.
41. Baskaran R, Wood LD, Whitaker LL et al. Ataxia telangiectasia mutant protein activates c-Abl tyrosine kinase in response to ionizing radiation. Nature 1997; 387:516-9.
42. Banin S, Moyal L, Shieh S et al. Enhanced phosphorylation of p53 by ATM in response to DNA damage. Science 1998; 281:1674-7.
43. Canman CE, Lim DS, Cimprich KA et al. Activation of the ATM kinase by ionizing radiation and phosphorylation of p53. Science 1998; 281:1677-9.
44. Khanna KK, Keating KE, Kozlov S et al. ATM associates with and phosphorylates p53: Mapping the region of interaction. Nat Genet 1998; 20:398-400.
45. Saito S, Goodarzi AA, Higashimoto Y et al. ATM mediates phosphorylation at multiple p53 sites, including Ser(46), in response to ionizing radiation. J Biol Chem 2002; 277:12491-4, Epub 2002 Mar 1.
46. Maya R, Balass M, Kim ST et al. ATM-dependent phosphorylation of Mdm2 on serine 395: Role in p53 activation by DNA damage. Genes Dev 2001; 15:1067-77.
47. Cortez D, Wang Y, Qin J et al. Requirement of ATM-dependent phosphorylation of brca1 in the DNA damage response to double-strand breaks. Science 1999; 286:1162-6.
48. Gatei M, Scott SP, Filippovitch I et al. Role for ATM in DNA damage-induced phosphorylation of BRCA1. Cancer Res 2000; 60:3299-304.
49. Xu B, O'Donnell AH, Kim ST et al. Phosphorylation of serine 1387 in Brca1 is specifically required for the Atm-mediated S-phase checkpoint after ionizing irradiation. Cancer Res 2002; 62:4588-91.
50. Li S, Ting NS, Zheng L et al. Functional link of BRCA1 and ataxia telangiectasia gene product in DNA damage response. Nature 2000; 406:210-5.

51. Lim DS, Kim ST, Xu B et al. ATM phosphorylates p95/nbs1 in an S-phase checkpoint pathway. Nature 2000; 404:613-7.
52. Bakkenist CJ, Kastan MB. DNA damage activates ATM through intermolecular autophosphorylation and dimer dissociation. Nature 2003; 421:499-506.
53. Horejsi Z, Falck J, Bakkenist CJ et al. Distinct functional domains of Nbs1 modulate the timing and magnitude of ATM activation after low doses of ionizing radiation. Oncogene 2004; 23:3122-7.
54. Kim ST, Xu B, Kastan MB. Involvement of the cohesin protein, Smc1, in Atm-dependent and independent responses to DNA damage. Genes Dev 2002; 16:560-70.
55. Yazdi PT, Wang Y, Zhao S et al. SMC1 is a downstream effector in the ATM/NBS1 branch of the human S-phase checkpoint. Genes Dev 2002; 16:571-82.
56. Matsuoka S, Rotman G, Ogawa A et al. Ataxia telangiectasia-mutated phosphorylates Chk2 in vivo and in vitro. Proc Natl Acad Sci USA 2000; 97:10389-94.
57. Ward IM, Wu X, Chen J. Threonine 68 of Chk2 is phosphorylated at sites of DNA strand breaks. J Biol Chem 2001; 276:47755-8, Epub 2001 Oct 19.
58. Brown AL, Lee CH, Schwarz JK et al. A human Cds1-related kinase that functions downstream of ATM protein in the cellular response to DNA damage. Proc Natl Acad Sci USA 1999; 96:3745-50.
59. Gatei M, Sloper K, Sorensen C et al. Ataxia-telangiectasia-mutated (ATM) and NBS1-dependent phosphorylation of Chk1 on Ser-317 in response to ionizing radiation. J Biol Chem 2003; 278:14806-11, Epub 2003 Feb 14.
60. Girard PM, Riballo E, Begg AC et al. Nbs1 promotes ATM dependent phosphorylation events including those required for G1/S arrest. Oncogene 2002; 21:4191-9.
61. Chen MJ, Lin YT, Lieberman HB et al. ATM-dependent phosphorylation of human Rad9 is required for ionizing radiation-induced checkpoint activation. J Biol Chem 2001; 276:16580-6, Epub 2001 Feb 6.
62. Vialard JE, Gilbert CS, Green CM et al. The budding yeast Rad9 checkpoint protein is subjected to Mec1/Tel1-dependent hyperphosphorylation and interacts with Rad53 after DNA damage. EMBO J 1998; 17:5679-88.
63. Kishi S, Zhou XZ, Ziv Y et al. Telomeric protein Pin2/TRF1 as an important ATM target in response to double strand DNA breaks. J Biol Chem 2001; 276:29282-91 Epub 2001 May 25.
64. Rogakou EP, Pilch DR, Orr AH et al. DNA double-stranded breaks induce histone H2AX phosphorylation on serine 139. J Biol Chem 1998; 273:5858-68.
65. Paull TT, Rogakou EP, Yamazaki V et al. A critical role for histone H2AX in recruitment of repair factors to nuclear foci after DNA damage. Curr Biol 2000; 10:886-95.
66. Bassing CH, Chua KF, Sekiguchi J et al. Increased ionizing radiation sensitivity and genomic instability in the absence of histone H2AX. Proc Natl Acad Sci USA 2002; 99:8173-8, Epub 2002 May 28.
67. Celeste A, Petersen S, Romanienko PJ et al. Genomic instability in mice lacking histone H2AX. Science 2002; 296:922-7, Epub 2002 April 4.
68. Andegeko Y, Moyal L, Mittelman L et al. Nuclear retention of ATM at sites of DNA double strand breaks. J Biol Chem 2001; 276:38224-30, Epub 2001 July 13.
69. Plug AW, Peters AH, Xu Y et al. ATM and RPA in meiotic chromosome synapsis and recombination. Nat Genet 1997; 17:457-61.
70. Brush GS, Morrow DM, Hieter P et al. The ATM homologue MEC1 is required for phosphorylation of replication protein A in yeast. Proc Natl Acad Sci USA 1996; 93:15075-80.
71. Shao RG, Cao CX, Zhang H et al. Replication-mediated DNA damage by camptothecin induces phosphorylation of RPA by DNA-dependent protein kinase and dissociates RPA: DNA-PK complexes. EMBO J 1999; 18:1397-406.
72. Wang H, Guan J, Perrault AR et al. Replication protein A2 phosphorylation after DNA damage by the coordinated action of ataxia telangiectasia-mutated and DNA-dependent protein kinase. Cancer Res 2001; 61:8554-63.
73. Oakley GG, Loberg LI, Yao J et al. UV-induced hyperphosphorylation of replication protein a depends on DNA replication and expression of ATM protein. Mol Biol Cell 2001; 12:1199-213.
74. Anderson L, Henderson C, Adachi Y. Phosphorylation and rapid relocalization of 53BP1 to nuclear foci upon DNA damage. Mol Cell Biol 2001; 21:1719-29.

75. Lee JH, Paull TT. Direct activation of the ATM protein kinase by the Mre11/Rad50/Nbs1 complex. Science 2004; 304:93-6.

76. Golding SE, Rosenberg E, Khalil A et al. Double strand break repair by homologous recombination is regulated by cell cycle-independent signaling via ATM in human glioma cells. J Biol Chem 2004; 279:15402-10, Epub 2004 Jan 26.

77. Takai H, Naka K, Okada Y et al. Chk2-deficient mice exhibit radioresistance and defective p53-mediated transcription. EMBO J 2002; 21:5195-205.

78. Rothkamm K, Kruger I, Thompson LH et al. Pathways of DNA double-strand break repair during the mammalian cell cycle. Mol Cell Biol 2003; 23:5706-15.

79. Foray N, Priestley A, Alsbeih G et al. Hypersensitivity of ataxia telangiectasia fibroblasts to ionizing radiation is associated with a repair deficiency of DNA double-strand breaks. Int J Radiat Biol 1997; 72:271-83.

80. Kuhne M, Riballo E, Rief N et al. A double-strand break repair defect in ATM-deficient cells contributes to radiosensitivity. Cancer Res 2004; 64:500-8.

81. Ribeiro JC, Barnetson AR, Jackson P et al. Caffeine-increased radiosensitivity is not dependent on a loss of G2/M arrest or apoptosis in bladder cancer cell lines. Int J Radiat Biol 1999; 75:481-92.

82. Wang X, Wang H, Iliakis G et al. Caffeine-induced radiosensitization is independent of nonhomologous end joining of DNA double-strand breaks. Radiat Res 2003; 159:426-32.

83. Wang H, Wang X, Iliakis G et al. Caffeine could not efficiently sensitize homologous recombination repair-deficient cells to ionizing radiation-induced killing. Radiat Res 2003; 159:420-5.

84. Barlow C, Dennery PA, Shigenaga MK et al. Loss of the ataxia-telangiectasia gene product causes oxidative damage in target organs. Proc Natl Acad Sci USA 1999; 96:9915-9.

85. Takao N, Li Y, Yamamoto K. Protective roles for ATM in cellular response to oxidative stress. FEBS Lett 2000; 472:133-6.

CHAPTER 8

Mechanisms of DNA Damage and Repair in Alzheimer Disease

V. Prakash Reddy, George Perry, Marcus S. Cooke, Lawrence M. Sayre and Mark A. Smith

Introduction

Reactive oxygen species (ROS) are produced during the respiratory cycle in mitochondria,[1] as well as normal cellular and xenobiotic metabolism. Exposure to various noxious insults can also lead to ROS production. In addition, ROS are also generated through metal-catalyzed reactions. A consequence of ROS production is the modification of cellular biomolecules, such as DNA, protein and lipids.[2,3] In addition to mutation, which is commonly considered, oxidative modification of DNA can have other, broad-ranging effects upon the function of the cell, impacting upon telomeres, microsatellite sequences, promoters and sites of methylation.[4,5] Perhaps as a consequence, such damage appears to have an important role in the pathogenesis of many diseases,[4] including Alzheimer disease (AD). Notably in AD, oxidative stress is regarded as one of the earliest pathological changes[6,7] and likely involves metabolic changes,[8] and redox-active metals[9,10] as well as other factors.[11] Oxidative modification of nuclear and mitochondrial DNA are thought to exacerbate AD, and their measurement is often used as a marker of oxidative stress. Indeed, extensive mitochondrial DNA damage was observed when PC12 cells were exposed to amyloid-β (Aβ), showing a direct correlation between oxidative stress and DNA damage.[12] Treatment of the exposed cells with endonuclease III or formamidopyrimidine (FaPy) glycosylase revealed significant damage to pyrimidine or purine bases, although in recent studies Aβ was found to sequester ROS, thus acting as an antioxidant equivalent.[13]

Nucleic acids may also be damaged by reactive nitrogen species (RNS), leading to deamination of the nucleotide bases. The RNS-mediated damage of DNA is associated with upregulation of nitrotyrosine which is also used as a marker for DNA damage by RNS.[14] In addition, DNA is also damaged by advanced lipid peroxidation end products (ALEs), for example, trans-4-hydroxy-2-nonenal (HNE), through the formation of DNA-HNE adducts.[15]

It is essential that damage to DNA does not persist. Repair of damaged DNA is accomplished mostly through nucleotide- or base-excision pathways (NER and BER, respectively). Base excision repair occurs through the mediation of enzymes called glycosylases which function by cleaving the damaged bases, which will be subsequently transported into cerebrospinal fluid (CSF). These enzymes selectively cleave phosphodiester bonds 5' and 3' to damaged bases, producing free hydroxyl and phosphate groups at the respective termini.[16] The released oxidized nucleosides such as 8-hydroxy-2'-deoxyguanosine (8-OHdG) are then passed through the circulation into urine and eventually excreted.[17] The decreased levels of damaged deoxynucleosides in

DNA Repair and Human Disease, edited by Adayabalam S. Balajee. ©2006 Landes Bioscience and Springer Science+Business Media.

CSF from cases of AD, as compared to those of control cases, give an indication of the absence or failure of the repair enzymes in AD. The latter effect is reflected also in the accumulation of damaged bases in intact DNA strands. In this review, we will focus upon studies related to the effects of DNA damage in AD and neuronal response to repair the damaged DNA.

Free Radical Formation and Biomolecular Damage

Formation of ROS

Metal (Fe^{2+} or Cu^+) catalyzed reduction of molecular oxygen gives rise to the superoxide anion which can be protonated at pH below 6 to give the hydroperoxyl radical (HOO•), which in turn can be converted to hydrogen peroxide (H_2O_2) by metal catalyzed reduction followed by protonation. Superoxide dismutase (SOD) can also catalyze the transformation of superoxide radical anion into hydrogen peroxide. Nonenzymatic superoxide dismutation is very rapid and will occur in biological compartments when the concentration of SOD is low. However, this can generate singlet oxygen, which can damage DNA. The superoxide anion and hydrogen peroxide by themselves are not highly oxidizing species, but the metal ion (e.g., Fe^{2+} or Cu^+) catalyzed Fenton reaction of H_2O_2 results in the formation of the highly toxic hydroxyl (•OH) radical. The generalized reaction sequence for the hydroxyl radical formation is shown below (Fig. 1).

Formation of RNS

Similarly, RNS, especially peroxynitrite, can initiate DNA damage. Peroxynitrite is derived from the reaction of nitric oxide (a free radical) with superoxide radical anion. Rapid protonation of peroxynitrite anion in cells gives peroxynitrous acid (ONOOH), which is a good nitrating agent for tyrosine and tryptophan side chains in proteins. Peroxynitrous acid can also react with deoxyribose backbones and cause single- and double-strand breaks. Further, it can cause nitration or deamination of DNA bases. Guanine, for example, upon deamination gives xanthine, and adenine similarly gives hypoxanthine. The mispairing of these newly generated oxidized forms of nucleic acids can result in nuclear base transversion mutations. Similar deaminative damage may result from alkylperoxynitrites, themselves formed by the reaction of nitric oxide with alkylperoxy radicals (Fig. 2).

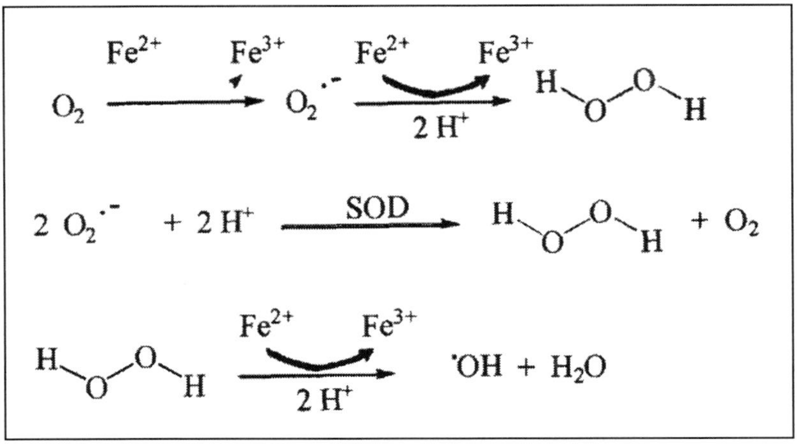

Figure 1. Intracellular formation of superoxide radical anions, hydrogen peroxide and hydroxyl radicals.

Figure 2. Formation of RNS, peroxynitrite and alkyl peroxynitrrite.

Advanced Glycation End Products and Advanced Lipid Peroxidation End Products

Adduction of reducing sugars to proteins and evolution of the initial adducts under aerobic conditions leads to advanced glycation end products (AGEs), which contribute to the histopathological and biochemical hallmarks of AD. Among such AGEs, pentosidine (a cross-link between lysine and arginine), pyrraline (a lysine-derived adduct), and carboxymethyllysine (CML, the product of condensation of glyoxal with lysine, which can also form from oxidative cleavage of a glycated lysine) are identified in elevated levels in AD brains.[18-23] Similarly, advanced lipid peroxidation end products (ALEs), formed by the reaction of proteins with lipid-peroxidation products, such as 4-hydroxy-*trans*-2-nonenal (HNE), are elevated in AD (Fig. 3). HNE derived adducts of proteins were shown to be present in AD by immunocytochemical studies, using antibodies specific to various HNE adducts.[24-26] The AGEs and ALEs can potentially act as further sources of ROS by their ability to chelate, and in some cases reduce, redox active transition metals. The AGEs also activate the receptor for AGE (RAGE), indirectly contributing to an increased production of ROS.[23,27]

DNA Damage Involving Lipoxidation-Derived Aldehydes

Lipid peroxidation generates numerous cytotoxic aldehydes. HNE, in particular, has been extensively explored in recent years in the pathogenesis of AD. It can form exocyclic

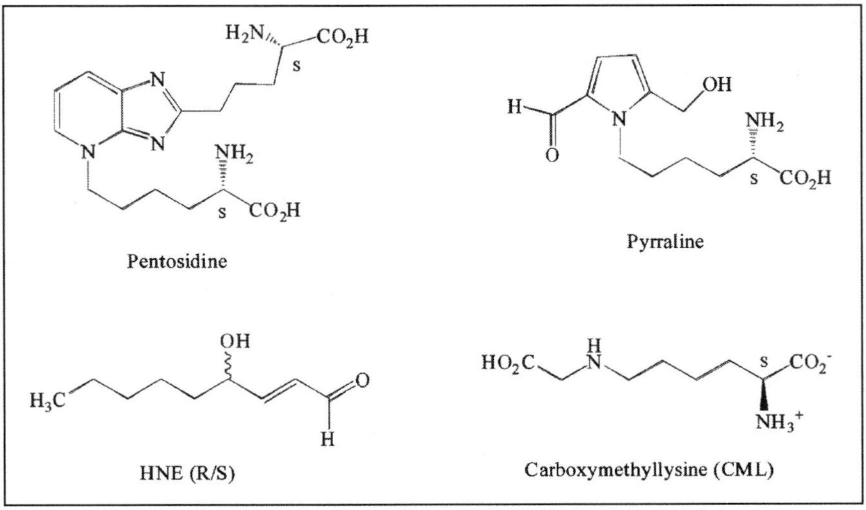

Figure 3. Structures of typical AGEs and HNE found in AD.

propano- adducts by reaction with nucleoside purine and pyrimidine bases. One such adduct involving the reaction of HNE with 2'-deoxyguanosine has been isolated.[28] The reaction presumably proceeds through the initial Michael addition of the free amino group (in the purine ring) to the HNE followed by nucleophilic addition of the N^1 (in the purine ring) to the carbonyl group. Subsequent dehydration gives the stable propano-adduct.

Among other products of lipid peroxidation, 4-oxo-2-nonenal (4-ONE) has been well characterized[29,30] and shown to be involved in DNA damage.[29,31-35] DNA forms exocyclic five-membered 'etheno-' adducts with 4-ONE through reactions with purine and pyrimidine rings. Several DNA/4-ONE adducts such as those derived from 2'-deoxyadenosine, 2'-deoxyguanosine, and 2'-deoxycytosine have been characterized by Blair and coworkers using atmospheric pressure ionization/MS/MS techniques[33] (Fig. 4). The formation of these etheno-adducts was postulated to involve initial nucleophilic addition of the free amino group

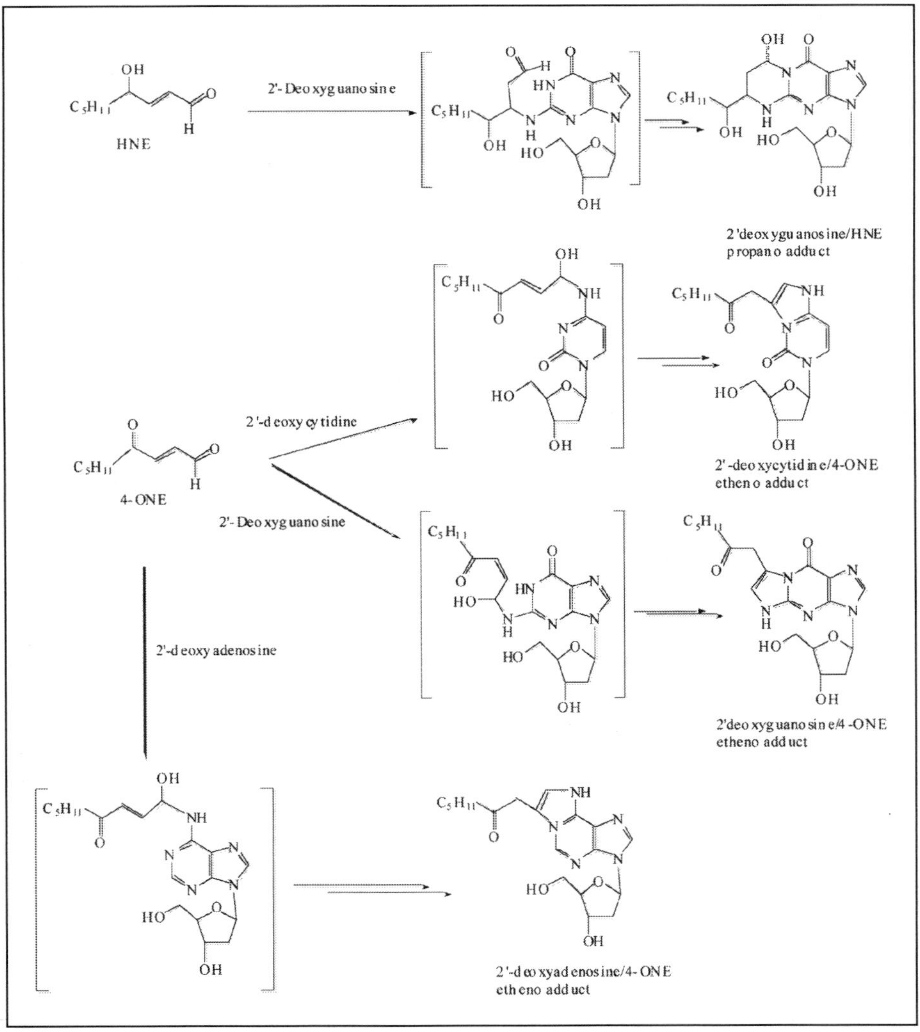

Figure 4. HNE and 4-ONE induced DNA damage.

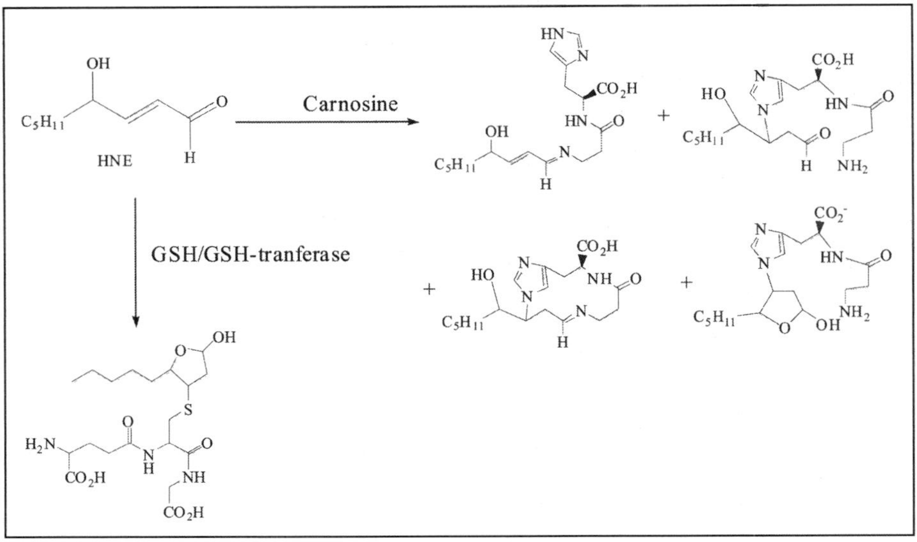

Figure 5. HNE detoxification by carnosine and glutathione (GSH).

of the purine/pyrimidine rings to the carbonyl group of 4-ONE, followed by Michael addition and subsequent dehydration reactions.[33] The etheno adducts are formed in high yields when 2'-deoxyguanosine and 2'-deoxyadenosine are treated with 4-ONE.[29]

4-ONE is more reactive and cytotoxic than HNE, and may be more biologically important. Its adduct with Vitamin C has been recently found in human plasma, verifying its generation in vivo.[36] Since 4-ONE also readily modifies proteins,[37] 4-ONE could interact with proteins responsible for nucleotide excision repair, damaging DNA repair mechanisms.

The effects of HNE on DNA damage and inhibition of DNA repair has been explored in human cells in the case of mutations responsible for cancer,[38] although there has been no such evidence of HNE modification of DNA repair enzymes in AD. In the absence of appropriate DNA repair mechanisms, the DNA adducts can induce apoptosis,[39] an important event in AD.

HNE Detoxification

HNE can be detoxified by anti-inflammatory agents such as carnosine through Michael adduct formation.[40,41] The glutathione transferase superfamily of enzymes catalyzes the nucleophilic attack of glutathione on HNE for detoxification (Fig. 5).[42]

Free Radical Formation from Amyloid-β (Aβ)

Aβ may also serve as a source of ROS, as it has been shown to bind to Cu^{2+} and Zn^{2+}, inducing its aggregation. However, it has been shown that copper binding to Aβ results in conformational changes that enable SOD-like or Cu/Zn SOD activity.[43] Thus Aβ functions to maintain oxidative balance. Further, it can act as an antioxidant by sequestering free radicals.[44] For this reason therapeutic intervention aimed at disrupting the relationship between heavy metals and Aβ is in doubt.[13]

Mechanisms of DNA Damage by ROS/RNS

ROS and RNS can damage DNA by modifying nucleic acid bases. Most of the damage of the nucleic acids arises directly or indirectly from the •OH radical produced from the metal catalyzed Fenton reaction.[45] ROS, for example the hydroxyl radical, can modify guanine, forming

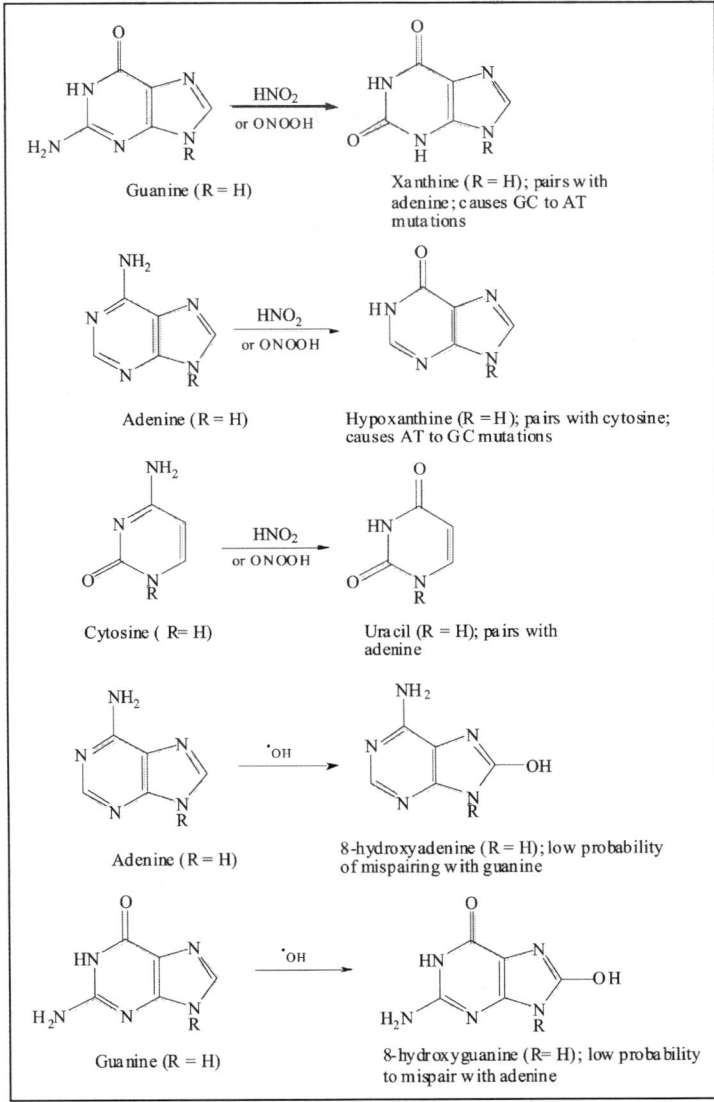

Figure 6. RNS and ROS induced oxidative deamination of nucleic acid bases.

8-hydroxyguanine, which energetically base pairs more favorably with adenine rather than with cytosine, the resulting mispairing producing GC to TA transversions.[46] Thymine similarly gives rise to 5-hydroxymethyluracil, as a common (but not sole) product upon reaction with ROS. It has a low probability of mispairing with guanine.[47] 8-Hydroxyadenine, a ROS derived product of adenine, also mispairs with guanine.[47]

Reaction of RNS (e.g., HNO_2 or ONOOH) with nucleic acid bases normally results in oxidative deamination, replacing NH_2 groups with OH groups. Thus adenine, cytosine and guanine are transformed into hypoxanthine, uracil, and xanthine, respectively (Fig. 6). Hypoxanthine mispairs with cytosine, and uracil mispairs with adenine. Deamination of nucleic acid bases can therefore result in AT to GC transversions.[47] A variety of oxidatively modified nucleic acid bases have been identified and measured by HPLC-MS/MS techniques. However, artifactual DNA damage during isolation and sample preparation makes the latter technique somewhat unreliable, and a number of work-up procedures to minimize such potential alterations have been introduced.[48] Immunocytochemical studies, using antibodies specific to known DNA modifications, are useful as they circumvent many of the problems associated with DNA extraction.[49]

Protection against ROS and RNS

Antioxidants, such as ascorbic acid, vitamin E, glutathione, and β-carotene can intercept superoxide and hydroperoxide radical anions as well as hydroxyl radicals, thereby limiting their toxic effects. In general, combinations of vitamin C and vitamin E are more effective as antioxidants compared to either one alone. In these cases, vitamin E can be regenerated from its oxidatively modified form by reaction with vitamin C, which is relatively more abundant in cells. CuZn superoxide dismutase and Mn superoxide dismutase react with superoxide radical anions and convert them into hydrogen peroxide and molecular oxygen. Downregulation of these enzymes can result in exacerbation of oxidative stress. However, in AD brains these enzymes are abundant, ruling out the possibility that the deficiency of these enzymes results in the pathogenesis of AD.[50]

Markers of DNA Damage in AD

Interaction of the hydroxyl radical with purine and pyrimidine bases can give rise to multiple products[51] (Fig. 7). Adenine and guanine can react with hydroxyl radicals at C_4, C_5, or C_8 positions in the pyrimidine ring. Thymine can undergo the addition of hydroxyl radicals at C_4 or C_5 positions to give the corresponding radicals, which upon reaction with O_2 give the peroxide products. Reductive cleavage of the latter peroxides can give *cis-* and *trans*-thymine glycols. Thymine can also suffer hydrogen abstraction from the methyl group to give the corresponding free radical, which upon reaction with O_2 followed by reductive cleavage gives the

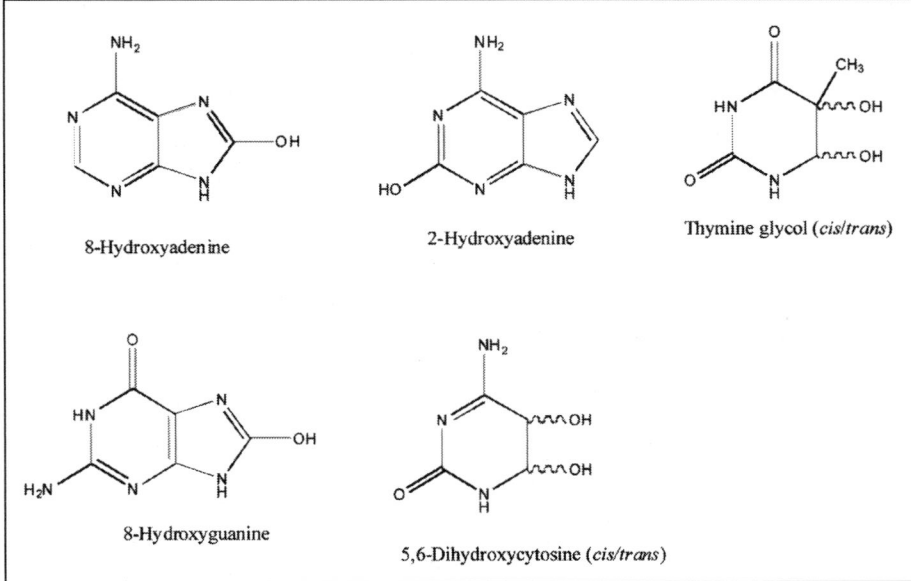

Figure 7. Structures of illustrative ROS generated nucleic acid bases.

corresponding alcohol. Cytosine similarly can give several products including cytosine glycol and 5,6-dihydroxycytosine. Thus although over 20 different oxidation products of DNA have been characterized, most of the investigators have focused attention on 8-hydroxyguanine, or its deoxynucleoside equivalent, 8-OHdG, a major oxidation product of guanine. Although the mechanism of the ring-opening of the imidazolinone ring to give the formamido derivative (Fapy-guanine) has not been established, it can be presumed to involve electron transfer followed by protonation at the carbonyl carbon and oxygen atoms (Fig. 8). 8-OHdG and Fapy-guanine were shown to be the major products of DNA oxidation by radiation-generated ROS and RNS.[52] Using gas chromatography/mass spectrometry techniques, Halliwell and coworkers identified the oxidized products of all four DNA bases: 8-hydroxyadenine, 8-hydroxyguanine, thymine glycol, Fapy-guanine, 5-hydroxyuracil, and Fapy-adenine in parietal, temporal, occipital, and frontal lobe, superior temporal gyrus, and hippocampus of AD brains,[53] implying a role for such DNA damage in the pathogenesis of AD.

Mitochondrial DNA (mtDNA) is more prone to oxidative damage as compared to that of nuclear DNA (nDNA), as the latter is well protected by histones and DNA-binding proteins. The mtDNA is also exposed to increased concentrations of ROS. Elevated levels of 8-OHdG, 8-hydroxyadenine, and 5-hydroxyuracil, a product of cytosine oxidation, were observed from the DNA of parietal, temporal and frontal lobes of AD brains as compared with age-matched controls,[54] although there was no significant correlation between the oxidized bases and

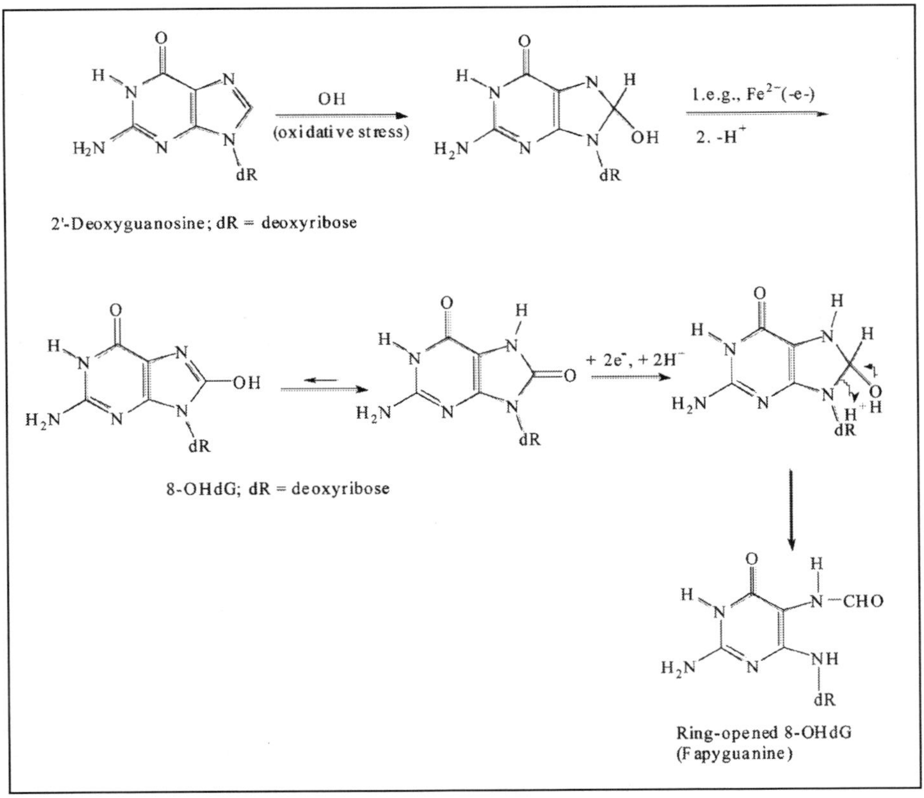

Figure 8. Oxidative stress of DNA leading to the formation of 8-OHdG, and its ring-opened formanido pyrimidine derivative.

neurofibrillary tangles or senile plaques. The levels of 8-OHdG were significantly elevated in intact DNA isolated from ventricular CSF of AD subjects, whereas the levels of the free nucleotide, formed due to the action of excision repair, were significantly lower in CSF.[55] These observations show that DNA damage is not only greater in AD, but also DNA repair mechanisms may be impaired in AD.

8-OHdG is present in significant amounts (about three-fold increase as compared to control cases) in the mtDNA and nDNA of AD brains.[56] Damage to mtDNA was significantly higher than that of nDNA. Lovell and coworkers have found statistically significant decreases in 8-OHdG glycosylase activity in the nuclear fraction of AD hippocampal and parahippocampal gyri (HPG), superior and middle temporal gyri (SMTG), and inferior parietal lobule (IPL). Depletion of helicase activity was also observed in the nuclear fraction in the IPL of AD.[57] Thus it is likely that the decreased repair of DNA damage could be one hallmark of the pathogenesis of AD. Furthermore, using the Comet assay with oxidative lesion-specific DNA repair endonucleases (endonuclease III for oxidized pyrimidines, Fapy-glycosylase for oxidized purines) it was determined that AD is associated with elevated levels of 8-OHdG and other oxidized nucleic acids.[58]

8-Hydroxyguanosine (8OHG)

In situ approaches involving immunochemical techniques are better suited for accurate localization of the 8-OHdG and 8OHG within cells and tissues. Smith and coworkers investigated the formation of these products in AD using immunoreactivity with monoclonal antibodies 1F7 and 15A3 specific to 8-OHdG and 8OHG.[59] From these studies it was found that 8-OHdG and 8OHG were prominent in the cytoplasm, and to a lesser extent in the nucleolus and nuclear envelope in neurons within the hippocampus, subiculum, and entorhinal cortex as well as frontal, temporal, and occipital neocortex in AD. Control cases were only faintly immunolabeled in similar structures. Damage to RNA is likely to be more extensive than to DNA, as there are more non-base-paired regions in RNA as compared to DNA, there are no protective histones, and the potential for repair has received only limited study.[4] Furthermore, it is plausible that the metals bound to RNA are major sites of redox activity, which result in the formation of hydroxyl radicals.

The oxidized nucleosides in AD are associated predominantly with RNA, as immunoreactivity towards 8OHG was diminished greatly by pre-incubation with RNase but only slightly by DNase. From these observations it can be inferred that mitochondria may be a major source of ROS that cause oxidative damage to DNA and RNA in AD.[6] A recent paper has suggested that mitochondrial ROS damage mitochondrial DNA, but not nuclear DNA.[60] It was concluded that the ROS derived from the mitochondrial respiratory chain are detoxified by mitochondrial SOD and other agents such as cytosolic catalase, and thus are not able to travel the distance from the mitochondria to the nucleus.

Quantitatively, neuronal 8OHG is greatest early in AD and is reduced with disease progression. Interestingly, nitrotyrosine, a marker of RNS mediated oxidative stress, is also elevated in the early stages of AD and is reduced with disease progression, indicating the significance of RNS in RNA damage in AD (vide infra). Smith and coworkers observed that the increase in Aβ deposition is associated with decreased oxidative damage.[7] Furthermore, neurons with neurofibrillary tangles show a 40% to 56% decrease in 8OHG levels compared with neurons free of neurofibrillary tangles, demonstrating that oxidative stress-induced RNA and DNA damage is an early event in AD, that decreases with disease progression.[7] A marked accumulation of 8OHG and nitrotyrosine, was also observed in the cytoplasm of cerebral neurons in Down's syndrome (DS), with the levels of nucleic acid and protein oxidation paralleling each other.[6] Thus in AD, increased levels of oxidative damage to DNA occur prior to the onset of Aβ deposition.

Mitochondrial versus Nuclear DNA Damage

Impaired mitochondrial function may result in accelerated DNA damage. The aging human brain accumulates substantial oxidative damage to mitochondrial DNA, which may be due to impaired mitochondrial function resulting in the increase of ROS, or by reducing ATP required for DNA repair.[61] The levels of 8-OHdG in DNA isolated from three regions of cerebral cortex and cerebellum increase progressively with normal aging in both nDNA and mtDNA.[62] The rate of increase of 8-OHdG, however, was 10-15 times higher in mtDNA than nDNA.[62] A significant reduction of mitochondrial fluidity was observed in AD along with increased levels of 8-OHdG in mtDNA. The alteration in membrane fluidity is primarily a result of lipid peroxidation. HNE and malondialdehyde (MDA), two widely studied lipid peroxidation products, were isolated in mitochondria, and more importantly, HNE was shown to modify the membrane fluidity by direct interaction with membrane phospholipids.[63] Thus, there is a direct correlation between oxidative stress and DNA damage, implicating oxidative stress in the pathogenesis of AD.[62]

The ROS and RNS generated as a result of impaired mitochondrial function, combined with reduced ATP levels in these mitochondria, can contribute to damage of vulnerable genes in the aging human brain.[61] The 8-OHdG may thus serve as a marker for oxidative stress in the aging brain. However, it is not clear whether it would have a significant effect on total cellular energy metabolism as there are thousands of mitochondria, it is unlikely that DNA damage may occur in the same gene in all of them.[64] However, when mutations occur in mtDNA, the corresponding mutant proteins may increase the inefficiency of the mitochondrial respiratory chain, resulting in the excessive production of ROS that cause further mtDNA and nDNA damage.

Neuronal DNA Damage by RNS

Smith and coworkers,[65] along with Su and coworkers,[14] have found evidence of peroxynitrite-mediated nitration of tyrosine residues of proteins to give 3-nitrotyrosine (NT)-derived proteins (*vide supra*). Using double labeling experiments with NT and PHF/ tau antibodies, these workers have found that the most intense NT-positive neurons usually contained neurofibrillary tangles. However, many neurons lacking neurofibrillary tangles are also intensely stained for NT. NT was undetectable in the cerebral cortex of age-matched control brains. Thus NT upregulation may precede neurofibrillary tangle formation. The peroxynitrous acid and other RNS such as nitrous acid (HONO) can deaminate guanine residues of DNA. Su and coworkers[14] showed that the terminal deoxynucleotidyl transferase (TdT)-labeled neurons are elevated in AD, and they have strong NT immunoreactivity, suggesting that the neurons with DNA damage in the absence of tangle formation may degenerate by tangle-independent mechanisms.

The neuronal form of nitric oxide synthase (nNOS) is elevated in reactive astrocytes in the hippocampus and entorhinal cortex in AD.[66] Although NO shows protective antioxidant effects, at high concentrations it is known to mediate DNA damage.[65] As a result of the upregulation of NO, increased cell death, related to DNA damage, was observed in the hippocampus and entorhinal cortex in AD.[66]

Diet Restriction and DNA Damage

DNA damage is significantly increased in senescent cells, e.g., in post-mitotic tissues of rodents of various ages.[8,67] A significant increase in 8-OHdG was observed in nDNA with increasing age in all tissues and strains of rodents studied, a result of increased sensitivity of these tissues to oxidative stress with age. Conversely, the age-dependent increase in 8-OHdG was reduced in the mtDNA of diet-restricted mice and rats.[67] It has been shown that ROS are reduced in diet-restricted rodents, where the rate of aging (e.g., life span extension) is also reduced.[68] Thus a major source of oxidative DNA damage is through ROS, which may be reduced through the use of antioxidants or metal chelators.

DNA Repair and AD

DNA damaged cells can undergo apoptosis through a common p53-dependent mechanism.[69] DNA damage can also be repaired by a combination of enzymes. If DNA damage is not repaired, it may be bypassed by specialized DNA polymerases.[70] The repair of DNA typically involves several steps. Excision enzymes such as DNA glycosylase remove the damaged bases by the cleavage of the base-sugar bonds. An endonuclease nicks the DNA strand at this site and removes it. Then, DNA synthase fills the missing link in the strand, and a DNA ligase joins this new DNA strand with the existing undamaged strand.[47] In addition to base excision repair, nucleotide excision repair may also be involved. The nucleotide excision repair involves removal of damaged nucleotides as part of large (30 nucleotide units) fragments.[71]

The accumulation of the damaged DNA bases in cells may result in the loss of normal cellular function which may be the basis of the causative factors of AD and other age related diseases. The oxidized bases can also be paired with abnormal bases during replication of DNA, leading, for example, to GG and AT combinations. In addition, pyrimidines next to the 8-OHdG residue are also misread[72] during DNA transcription.

Nucleotide/Base Excision Repair

Nucleotide excision repair requires the TFIIH transcription-repair complex having helicase activity. These DNA helicases unwind DNA for repair as well as for replication. Base excision repair, nucleotide excision repair, and mismatch repair have been identified and characterized in eukaryotes.[71] A variety of other protein complexes are also involved in DNA repair as outlined below.

There are a variety of base-specific glycosylases that cleave specific damaged bases.[73] For example, human HeLa cell extracts contain two forms of 8-hydroxyguanine glycosylase (hOGG) repair enzymes. hOGG-1 cleaves 8-hydroxyguanine paired with cytosine and thymine, whereas hOGG-2 cleaves 8-hydroxyguanine paired with adenine. Expression of DNA excision repair cross complementing proteins p80 and p89 was observed in AD brains.[74] These are known to repair different types of DNA damage.

It was found that the concentrations of 8-OHdG in CSF of AD are 100-fold higher than those of healthy individuals. In addition, the ratio of 8-OHdG to 8-hydroxyguanine is approximately 8-fold higher in CSF than in urine, suggesting that the nucleotide excision repair is a major DNA repair mechanism in removal of oxidatively damaged DNA in brain cells.[75]

Bcl-2 in DNA Repair

Bcl-2, a 26-kDa integral membrane protein, has an antioxidant effect and is increased in neurons with DNA damage in AD, as are other members of the Bcl family.[76,77] Expression of Bcl-2 in PC12 cells inhibit nitric oxide donor (sodium nitroprusside)- and peroxynitrite-induced cell death.[78] It was shown that H_2O_2, nitric oxide and peroxynitrite induced oxidative stress results in DNA damage both in mtDNA and nDNA, even in the presence of Bcl-2. However, recovery from DNA damage was accelerated in cells expressing Bcl-2, implicating that neuronal up-regulation of Bcl-2 may facilitate DNA repair after oxidative stress.[78] A growth arrest DNA damage-inducible protein, GADD45, is expressed in AD neurons and is associated with expression of Bcl-2.[79] GADD45 may aid in DNA repair. GADD45 was found to bind to PCNA, a normal component of Cdk complexes and a protein involved in DNA replication and repair, and stimulated DNA excision repair in vitro.[18]

Repair of double strand breaks requires DNA-dependent protein kinase, composed of DNA-PKcs and Ku. It was shown that Ku DNA binding activity was reduced in extracts of postmortem AD midfrontal cortex, and the decreased Ku DNA binding is correlated with reduced protein levels of Ku subunits (DNA-PKcs) and poly(ADP-ribose) polymerase-1.[80] Immunohistochemical analysis also suggested that DNA-PK protein levels reflected the number of neurons and regulation of cellular expression.

Repair of 8-OHdG

8-Oxoguanine DNA glycosylase (hOGG1-2a) is one of the excision repair enzymes that repairs 8-OHdG. Using an antibody specific to the mitochondrial form of hOGG1-2a, it was found that hOGG1-2a is expressed mainly in the neuronal cytoplasm in both AD and control cases in regionally different manners.[81] Immunoreactivity to hOGG1-2a is associated with neurofibrillary tangles, dystrophic neurites and reactive astrocytes in AD. The relatively low levels of expression of hOGG1-2a in AD indicates that oxidative DNA damage in mitochondria may be involved as the pathogenic factor.[81] Another repair enzyme for oxidative DNA damage, purine-nucleoside triphosphatase (hMTH1) was also observed in neurons of AD. In vitro studies showed that hOGG1-2a immunoreactivities in reactive astrocytes and oligodendrocytes were more intense than those to hMTH1.[82] By adding H_2O_2 to the cultured astrocyte cells, rapid induction of hMTH1 was observed, whereas the levels of hOGG1-2a were mildly increased, showing that hMTH1 is an inducible enzyme under oxidative stress, and hOGG1-2a is rather constitutively expressed and also up-regulated in the chronic stage of the disease.[82]

Mre11 DNA Repair Complex

The Mre11 protein complex consisting of Rad50, Mre11 and Nbs1 is essential for cellular responses to DNA damage, such as initiating cell cycle checkpoints and repairing damaged DNA. It was shown that the Mre11 complex proteins are present in neurons of adult human cortex and cerebellum and were substantially reduced in the neurons of AD cortex. The accumulated DNA damage in AD neurons may be, in part, as a result of the reduced levels of Mre11 protein complexes.[83]

Poly(ADP-Ribose) Polymerase (PARP)

Poly(ADP-ribose) polymerase (PARP) is one of the DNA repair enzymes that is activated as a result of oxidative stress induced single strand or double strand breaks of DNA. PARP catalyzes the cleavage of NAD^+ into adenosine 5'-diphosphoribose (ADP-ribose) and nicotinamide. It also catalyzes the covalent attachment of ADP-ribose polymers to nuclear proteins such as histones. PARP and poly(ADP-ribose)-immunolabelled neurons were detected in a much higher proportion in AD than in controls.[84] Overactivation of PARP causes massive NAD^+ depletion resulting in cell death due to energy depletion.

Conclusions

As highlighted above, the role of oxidative stress in the pathogenesis of AD is a burgeoning field. However, while much is known, much remains unknown as to the impact of therapeutic intervention in patients with disease. Translation of basic scientific findings into efficacious treatment strategies remains to be determined.

Acknowledgements

Work in the authors' laboratories is supported by the NIH (NS38648 and AG14249) and the Alzheimer's Association (IIRG-03-6263 and IIRG-04-1272).

References

1. Loft S, Poulsen HE. Cancer risk and oxidative DNA damage in man. J Mol Med 1996; 74;297-312.
2. Sayre LM, Smith MA, Perry G. Chemistry and biochemistry of oxidative stress in neurodegenerative disease. Curr Med Chem 2001; 8:721-738.
3. Perry G, Sayre LM, Atwood CS et al. The role of iron and copper in the aetiology of neurodegenerative disorders: Therapeutic implications. CNS Drugs 2002; 16:339-352.
4. Evans MD, Cooke MS. Factors contributing to the outcome of oxidative damage to nucleic acids. Bioessays 2004; 26:533-542.
5. Evans MD, Dizdaroglu M, Cooke MS. Oxidative DNA damage and disease: induction, repair and significance. Mutat Res 2004; 567:1-61.
6. Nunomura A, Perry G, Pappolla MA et al. Neuronal oxidative stress precedes amyloid-b deposition in Down syndrome. J Neuropathol Exp Neurol 2000; 59:1011-1017.
7. Nunomura A, Perry G, Aliev G et al. Oxidative damage is the earliest event in Alzheimer disease. J Neuropathol Exp Neurol 2001; 60:759-767.
8. Hosokawa M, Fujisawa H, Ax S et al. Age-associated DNA damage is accelerated in the senescence-accelerated mice. Mech Ageing Dev 2000; 118:61-70.
9. Smith MA, Harris PLR, Sayre LM et al. Iron accumulation in Alzheimer disease is a source of redox-generated free radicals. Proc Natl Acad Sci USA 1997; 94:9866-9868.
10. Sayre LM, Perry G, Harris PLR et al. In situ oxidative catalysis by neurofibrillary tangles and senile plaques in Alzheimer's disease: a central role for bound transition metals. J Neurochem 2000; 74:270-279.
11. Price DL, Rhett PM, Thorpe SR et al. Chelating activity of advanced glycation end-product inhibitors. J Biol Chem 2001; 276:48967-48972.
12. Bozner P, Grishko V, Ledoux SP et al. The amyloid-b protein induces oxidative damage of mitochondrial DNA. J Neuropathol Exp Neurol 1997; 56:1356-1362.
13. Smith MA, Joseph JA, Perry G et al. Tracking the culprit in Alzheimer's disease. Ann. N.Y. Acad Sci 2000; 924:35-38.
14. Su JH, Deng G, Cotman CW. Neuronal DNA damage precedes tangle formation and is associated with up-regulation of nitrotyrosine in Alzheimer's disease brain. Brain Res 1997; 774:193-199.
15. Gotz ME, Wacker M, Luckhaus C et al. Unaltered brain levels of $1,N_2$-propanodeoxyguanosine adducts of trans-4-hydroxy-2-nonenal in Alzheimer's disease. Neurosci Lett 2002; 324:49-52.
16. Chung MH, Kim HS, Ohtsuka E et al. An endonuclease activity in human polymorphonuclear neutrophils that removes 8-hydroxyguanine residues from DNA. Biochem Biophys Res Commun 1991; 178:1472-1478.
17. Shigenaga MK, Aboujaoude EN, Chen Q et al. Assays of oxidative DNA damage biomarkers 8-oxo-2'-deoxyguanosine and 8-oxoguanine in nuclear DNA and biological fluids by high-performance liquid chromatography with electrochemical detection. Methods Enzymol 1994; 234:16-33.
18. Smith ML, Chen IT, Zhan Q et al. Interaction of the p53-regulated protein Gadd45 with proliferating cell nuclear antigen. Science (Washington, D.C.) 1994; 266:1376-1380.
19. Vitek MP, Bhattacharya K, Glendening JM et al. Advanced glycation end products contribute to amyloidosis in Alzheimer disease. Proc Natl Acad Sci USA 1994; 91:4766-4770.
20. Yan SD, Chen X, Schmidt AM et al. Glycated tau protein in Alzheimer disease: a mechanism for induction of oxidant stress. Proc Natl Acad Sci USA 1994; 91:7787-7791.
21. Smith MA, Sayre LM, Monnier VM et al. Radical AGEing in Alzheimer's disease. Trends Neurosci 1995; 18:172-176.
22. Castellani RJ, Harris PLR, Sayre LM et al. Active glycation in neurofibrillary pathology of Alzheimer's disease: N-e-(carboxymethyl)lysine and hexitol-lysine. Free Radical Biol Med 2001; 31:175-180.
23. Reddy VP, Obrenovich ME, Atwood CS et al. Involvement of Maillard reactions in Alzheimer disease. Neurotoxicity Res 2001; 4:191-209.
24. Sayre LM, Zelasko DA, Harris PL et al. 4-Hydroxynonenal-derived advanced lipid peroxidation end products are increased in Alzheimer's disease. J Neurochem 1997; 68:2092-2097.
25. Takeda A, Smith MA, Avila J et al. In Alzheimer's disease, heme oxygenase is coincident with Alz50, an epitope of tau induced by 4-hydroxy-2-nonenal modification. J Neurochem 2000; 75:1234-1241.

26. Wataya T, Nunomura A, Smith Mark A et al. High molecular weight neurofilament proteins are physiological substrates of adduction by the lipid peroxidation product hydroxynonenal. J Biol Chem 2002; 277:4644-4648.
27. Munch G, Schinze R, Loske C et al. Alzheimer's disease - synergistic effects of glucose deficit, oxidative stress and advanced glycation endproducts. J Neural Transm 1998; 105:439-461.
28. Burcham PC. Genotoxic lipid peroxidation products: their DNA-damaging properties and role in formation of endogenous DNA adducts. Mutagenesis 1998; 13:287-305.
29. Lee SH, Blair IA. Characterization of 4-Oxo-2-nonenal as a Novel Product of Lipid Peroxidation. Chem Res Toxicol 2000; 13:698-702.
30. Spiteller P, Kern W, Reiner J et al. Aldehydic lipid peroxidation products derived from linoleic acid. Biochim Biophys Acta 2001; 1531:188-208.
31. Rindgen D, Nakajima M, Wehrli S et al. Covalent modifications to 2'-deoxyguanosine by 4-oxo-2-nonenal, a novel product of lipid peroxidation. Chem Res Toxicol 1999; 12;1195-1204.
32. Rindgen D, Lee SH, Nakajima M et al. Formation of a substituted 1,N6-etheno-2'-deoxyadenosine adduct by lipid hydroperoxide-mediated generation of 4-oxo-2-nonenal. Chem Res Toxicol 2000; 13:846-852.
33. Blair IA. Lipid hydroperoxide-mediated DNA damage. Exp Gerontol 2001; 36:1473-1481.
34. Pollack M, Oe T, Lee SH et al. Characterization of 2'-deoxycytidine adducts derived from 4-oxo-2-nonenal, a novel Lipid peroxidation product. Chem Res Toxicol 2003; 16:893-900.
35. Kawai Y, Uchida K, Osawa T. 2'-Deoxycytidine in free nucleosides and double-stranded DNA as the major target of lipid peroxidation products. Free Radical Biol Med 2004 36:529-541.
36. Sowell J, Frei B, Stevens JF. Vitamin C conjugates of genotoxic lipid peroxidation products: Structural characterization and detection in human plasma. Proc Natl Acad Sci USA 2004; 101:17964-17969.
37. Zhang WH, Liu J, Xu G et al. Model studies on protein side chain modification by 4-oxo-2-nonenal. Chem Res Toxicol 2003; 16:512-523.
38. Feng Z, Hu W, Tang MS et al. Trans-4-hydroxy-2-nonenal inhibits nucleotide excision repair in human cells: A possible mechanism for lipid peroxidation-induced carcinogenesis. Proc Natl Acad Sci USA 2004; 101:8598-8602.
39. West JD, Ji C, Duncan ST et al. Induction of Apoptosis in Colorectal Carcinoma Cells Treated with 4-Hydroxy-2-nonenal and Structurally Related Aldehydic Products of Lipid Peroxidation. Chem Res Toxicol 2004; 17:453-462.
40. Aldini G, Carini M, Beretta G et al. Carnosine is a quencher of 4-hydroxy-nonenal: through what mechanism of reaction?, Biochem Biophys Res Commun 2002; 298:699-706.
41. Aldini G, Granata P, Carini M et al. Detoxification of cytotoxic a,b-unsaturated aldehydes by carnosine: characterization of conjugated adducts by electrospray ionization tandem mass spectrometry and detection by liquid chromatography/mass spectrometry in rat skeletal muscle. J Mass Spectrom 2002; 37:1219-1228.
42. Xie C, Lovell MA, Markesbery WR. Glutathione transferase protects neuronal cultures against 4-hydroxynonenal toxicity. Free Radical Biol Med 1998 25:979-988.
43. Curtain CC, Ali F, Volitakis I et al. Alzheimer's disease amyloid-b binds copper and zinc to generate an allosterically ordered membrane-penetrating structure containing superoxide dismutase-like subunits. J Biol Chem 2001; 276:20466-20473.
44. Rottkamp CA, Raina AK, Zhu X et al. Redox-active iron mediates amyloid-b toxicity. Free Radical Biol Med 2001; 30:447-450.
45. Castellani RJ, Honda K, Zhu X et al. Contribution of redox-active iron and copper to oxidative damage in Alzheimer disease. Ageing Res Rev 2004; 3:319-326.
46. Shibutani S, Takeshita M, Grollman AP. Insertion of specific bases during DNA synthesis past the oxidation-damaged base 8-oxodG. Nature 1991; 349:431-434.
47. Halliwell B, Gutteridige JMC. Free radicals in biology and medicine. 3rd ed. Oxford University Press: New York, 1999.
48. Douki T, Ravanat JL, Frelon S et al. HPLC-MS/MS measurement of oxidative base damage to isolated and cellular DNA. Critical Reviews of Oxidative Stress and Aging 2003; 1:190-202.
49. Cooke MS, Lunec J. Critical Reviews of Oxidative Stress and Aging. In: Cutler RG, Rodriguez H, eds. New York: World Scientific Publishing, 2003; 1:275-293.

50. Marklund SL, Adolfsson R, Gottfries CG et al. Superoxide dismutase isoenzymes in normal brains and in brains from patients with dementia of Alzheimer type. J Neurol Sci 1985; 67:319-325.
51. Breen AP, Murphy JA. Reactions of oxyl radicals with DNA. Free Radical Biol Med 1995; 18:1033-1077.
52. Gajewski E, Rao G, Nackerdien Z et al. Modification of DNA bases in mammalian chromatin by radiation-generated free radicals. Biochemistry 1990; 29:7876-7882.
53. Lyras L, Cairns NJ, Jenner A et al. An assessment of oxidative damage to proteins, lipids, and DNA in brain from patients with Alzheimer's disease. J Neurochem 1997; 68:2061-2069.
54. Gabbita SP, Lovell MA, Markesbery WR. Increased nuclear DNA oxidation in the brain in Alzheimer's disease. J Neurochem 1998; 71:2034-2040.
55. Lovell MA, Gabbita SP, Markesbery WR. Increased DNA oxidation and decreased levels of repair products in Alzheimer's disease ventricular CSF. J Neurochem 1999; 72:771-776.
56. Mecocci P, MacGarvey U, Beal MF. Oxidative damage to mitochondrial DNA is increased in Alzheimer's disease. Ann Neurol 1994; 36:747-751.
57. Lovell MA, Xie C, Markesbery WR. Decreased base excision repair and increased helicase activity in Alzheimer's disease brain. Brain Res 2000; 855:116-123.
58. Kadioglu E, Sardas S, Aslan S et al. Esat Karakaya, Detection of oxidative DNA damage in lymphocytes of patients with Alzheimer's disease. Biomarkers 2004; 9:03-209.
59. Nunomura A, Perry G, Pappolla MA et al. RNA oxidation is a prominent feature of vulnerable neurons in Alzheimer's disease. J Neurosci 1999; 19:1959-1964.
60. Hoffmann S, Spitkovsky D, Radicella JP et al. Wiesner, Reactive oxygen species derived from the mitochondrial respiratory chain are not responsible for the basal levels of oxidative base modifications observed in nuclear DNA of mammalian cells. Free Radical Biol Med 2004; 36:765-773.
61. Lu T, Pan Y, Kao SY et al. Gene regulation and DNA damage in the ageing human brain. Nature 2004; 429:883-891.
62. Mecocci P, MacGarvey U, Kaufman AE et al. Oxidative damage to mitochondrial DNA shows marked age-dependent increases in human brain. Ann Neurol 1993; 34:609-616.
63. Chen JJ, Yu BP. Alterations in mitochondrial membrane fluidity by lipid peroxidation products. Free Radical Biol Med 1994; 17:411-418.
64. Kanaar R, Hoeijmakers JHJ. Recombination and joining: different means to the same ends. Genes Funct 1997; 1:165-174.
65. Smith MA, Harris PLR, Sayre LM et al. Widespread peroxynitrite-mediated damage in Alzheimer's disease. J Neurosci 1997; 17:2653-2657.
66. Simic G, Lucassen PJ, Krsnik Z et al. Bogdanovi, nNOS expression in reactive astrocytes correlates with increased cell death related DNA damage in the hippocampus and entorhinal cortex in Alzheimer's disease. Exp Neurol 2000; 165:12-26.
67. Hamilton ML, Van Remmen H, Drake JA et al. Does oxidative damage to DNA increase with age? Proc Natl Acad Sci USA 2001; 98: 10469-10474.
68. Sohal RS, Mockett RJ, Orr WC. Mechanisms of aging: An appraisal of the oxidative stress hypothesis. Free Radical Biol Med 2002; 33:575-586.
69. Itahana K, Dimri G, Campisi J. Regulation of cellular senescence by p53. Eur J Biochem 2001; 268:2784-2791.
70. Lindahl T, Wood RD. Quality control by DNA repair. Science (Washington, D. C.) 1999; 286:1897-1905.
71. Friedberg EC, Wood RD. DNA excision repair pathways. DNA Replication in Eukaryotic Cells, Cold Spring Harbor Monograph Series. DNA Replication in Eukaryotic Cells 1996; 31:249-269.
72. Kuchino Y, Mori F, Kasai H et al. Misreading of DNA templates containing 8-hydroxydeoxyguanosine at the modified base and at adjacent residues. Nature 1987; 327:77-79.
73. Hazra TK, Izumi T, Maidt L et al. The presence of two distinct 8-oxoguanine repair enzymes in human cells: their potential complementary roles in preventing mutation. Nucleic Acids Res 1998; 26:5116-5122.
74. Hermon M, Cairns N, Egly JM et al. Expression of DNA excision-repair-cross-complementing proteins p80 and p89 in brain of patients with Down Syndrome and Alzheimer's disease. Neurosci Lett 1998; 251:45-48.

75. Rozalski R, Winkler P; Gackowski D et al. High concentrations of excised oxidative DNA lesions in human cerebrospinal fluid. Clin Chem (Washington, DC, U.S.) 2003; 49:1218-1221.

76. Su JH, Anderson AJ, Cummings BJ et al. Immunohistochemical evidence for apoptosis in Alzheimer's disease. Neuroreport 1994; 5:529-2533.

77. Zhu X, Wang Y, Ogawa O et al. Neuroprotective properties of Bcl-w in Alzheimer disease. J Neurochem 2004; 89:1233-1240.

78. Deng G, Su JH, Ivins KJ et al. Bcl-2 facilitates recovery from DNA damage after oxidative stress. Exp Neurol 1999; 159:309-318.

79. Torp R, Su JH, Deng G et al. GADD45 is induced in Alzheimer's disease, and protects against apoptosis in vitro. Neurobiol Dis 1998; 5:245-252.

80. Davydov V, Hansen LA, Shackelford DA et al. Is DNA repair compromised in Alzheimer's disease? Neurobiol Aging 2003; 24:953-968.

81. Iida T, Furuta A, Nishioka K et al. Expression of 8-oxoguanine DNA glycosylase is reduced and associated with neurofibrillary tangles in Alzheimer's disease brain. Acta Neuropathol (Berl) 2002; 103:20-25.

82. Iida T, Furuta A, Nakabeppu Y et al. Defense mechanism to oxidative DNA damage in glial cells. Neuropathology 2004; 24:125-130.

83. Jacobsen E, Beach T, Shen Y et al. Deficiency of the Mre11 DNA repair complex in Alzheimer's disease brains. Mol Brain Res 2004; 128:1-7.

84. Love S, Barber R, Wilcock GK. Increased poly(ADP-ribosyl)ation of nuclear proteins in Alzheimer's disease. Brain 1999; 122:247-253.

CHAPTER 9

Orchestration of Telomeres and DNA Repair Factors in Mammalian Cells:
Implications for Cancer and Ageing

M. Prakash Hande

Abstract

L oss of telomere homeostasis via chromosome-genomic instability might effectively promote tumour progression. Telomere function may have contrasting roles: inducing replicative senescence and promoting tumourigenesis and these roles may vary between cell types depending on the expression of telomerase enzyme, the level of mutations induced, and deficiency of related DNA repair pathways. Earlier studies in yeast and their recent extension to mammalian systems have convincingly indicated a role for DNA repair proteins in telomere maintenance. An alternative telomere maintenance mechanism has been identified in mouse embryonic stem cells lacking the telomerase RNA unit (mTERC) in which nontelomeric sequences adjacent to existing short stretches of telomere repeats are amplified. Our quest for identifying telomerase-independent or alternative mechanisms for telomere maintenance in mammalian cells has identified the involvement of potential DNA repair factors in such pathways. Studies by us and others have shown the association between the DNA repair factors and telomere function in mammalian cells. Mice deficient in a DNA-break sensing molecule, PARP-1 (poly [ADP]-ribopolymerase), have increased levels of chromosomal instability associated with extensive telomere shortening. Ku80 null cells showed telomere shortening associated with extensive chromosome end fusions whereas Ku80$^{+/-}$ cells exhibited an intermediate level of telomere shortening. This overview will focus mainly on the role of DNA repair/recombination and DNA damage signalling molecules such as PARP-1, DNA-PKcs, Ku70/80, XRCC4 and ATM, which we have been studying for quite sometime. As the maintenance of telomere function is crucial for genomic stability, our results are likely to provide new insights into the telomere regulatory mechanisms and their impact on chromosome instability, ageing and tumour formation.

Telomeres and Telomerase

Telomeres are situated at the ends of linear chromosomes. In most eukaryotes, telomeres consist of non-coding (TTAGGG)$_n$ repeats that are associated with an array of proteins.[1] Vertebrate telomeres contain such repeated sequences thought to be folded by telomere binding proteins into a duplex T loop structure.[2,3] They cap the chromosomes to prevent DNA repair

---

DNA Repair and Human Disease, edited by Adayabalam S. Balajee. ©2006 Landes Bioscience and Springer Science+Business Media.

pathways to be triggered for the repair of spontaneous or induced breaks, [4-6] which may result in chromosomal end-to-end fusion. Fusions can be induced through perturbation of the telomeric DNA sequence or depletion of telomere-associated proteins, such that the fundamental structure of the telomere is altered or telomere homeostasis is lost. Specific DNA-binding proteins recognise and associate with either double-strand or single-strand telomeric repeat sequences providing further protection for the chromosome termini.

Telomere repeats are synthesised by the reverse transcriptase enzyme telomerase. [7,8] Telomerase is required for the stable maintenance of telomere repeats in vivo and in vitro. [9] The appropriate complement of telomere-associated proteins must be present to ensure that the telomere assumes a fully capped configuration and is excluded from processing by DNA damage repair systems. [10] Telomeres are believed to serve a protective function in the maintenance of chromosomal genome stability. [11] Chromosomes with defective telomere length and/ or structure are more susceptible in forming end-to-end fusions that fail to segregate properly in mitosis, resulting in chromosomal breakage accumulation and DNA damage checkpoint activation. The addition of telomeric tracts by telomerase compensates for the progressive loss of telomeric sequences from successive rounds of DNA replication.

Telomeres and Ageing

In most somatic cells, telomeric DNA is lost every time a cell divides. [12,13] As a result of this progressive telomere shortening, somatic cells cease to proliferate and become senescent after finite divisions (~60 population doublings). [14] Telomere shortening occurs rapidly in certain cell lines derived from premature ageing disorders, i.e., Werner Syndrome (WS) and Ataxia telangiectasia (AT) leading to premature senescence as compared to age-matched control cell lines [15-17] Additionally, cells derived from AT patients show chromosome abnormalities in the form of end-to-end fusions involving telomeres One characteristic feature of Werner Syndrome cells is the "variegated translocation mosaicism" which could be due to shortened telomeres. Thus, telomere shortening is directly related to ageing and senescence in in vitro model systems. However, a recent study has proposed that telomere dynamics at the single cell level in WS fibroblasts are not significantly different from those in normal fibroblasts, and suggest that the accelerated replicative decline seen in WS fibroblasts does not result from accelerated telomere erosion. [18]

Telomeres and Cancer

The loss of telomeres with age and the resulting genomic instability coupled with onset of telomerase activity are of critical importance in tumourigenesis and in tumour progression. [19] Telomere shortening in somatic cells and the subsequent replicative senescence may prevent uncontrolled cell proliferation and thereby malignant transformation. The tumour suppressor function of telomeres has also been suggested. [19,20] However, extensive proliferation and the subsequent telomere shortening/lengthening may also result in telomere dysfunction. Chromosome fusions and breaks resulting from telomere dysfunction may facilitate the loss of heterozygosity (LOH) of tumour suppressor genes. Preferential loss of telomere repeats from the ends of a particular chromosome harbouring tumour suppressor genes or oncogenes may result either in their inactivation or activation, repsectively, due to chromosome translocation events and may also facilitate tumour progression. Collectively, telomere dysfunction (see later) and associated chromosome/ genetic instability might effectively promote tumour progression. Thus telomere function may have contrasting roles: inducing replicative senescence and promoting tumourigenesis and these roles may vary between cell types depending on the expression of the enzyme telomerase, the level of mutations induced, and efficiency/deficiency of related DNA repair pathways.

Telomeres and DNA Repair Factors

The question of whether or not telomere shortening triggers double strand DNA break response is still largely unanswered. Telomerase knockout mice showed progressive shortening of telomeres up to the 6th generation after which they became infertile.[9,21] However, embryonic fibroblasts obtained from mTER[-/-] (mouse telomerase RNA or mTERC; telomerase negative) mice could be immortalized in vitro and displayed telomere maintenance with associated chromosomal defects at later passages in culture.[22] This study has implicated the existence of telomerase-independent mechanisms in the telomere maintenance. Similarly, we have identified an alternative telomere maintenance mechanism in mouse embryonic stem cells lacking telomerase RNA unit (mTER) with amplification of nontelomeric sequences adjacent to existing short stretches of telomere repeats.[23] Recombination of sub-telomeric sequences has been implicated in the telomere maintenance mechanisms in the telomerase negative mouse embryonic stem cells.[23] Our quest for identifying telomerase-independent or alternative mechanisms involved in telomere maintenance has implicated the involvement of potential DNA repair factors in such pathways. Several studies have shown the association between the DNA repair factors and telomere function. Extensive studies in yeast have linked the role of DNA repair/recombination and damage signalling molecules in telomere maintenance mechanisms. However, only recently, such roles for these proteins in mammalian cells have been uncovered. This overview will focus mainly on the role of DNA repair /recombination and DNA damage signalling molecules such as DNA-PKcs, ATM, Ku complex, XRCC4 and PARP in telomere chromosome integrity in mammalian cells for which we have sufficient knowledge and data.

SCID (Severe Combined Immunodeficiency) and Telomeres

In one of our earlier studies, perhaps one of the first in mammalian models, we have shown that mouse *scid* cells possess abnormal telomere lengths[24] being ten times longer than their CB17 parental cells. Besides having abnormally long telomeres, the *scid* cell line showed unusual telomeric associations. Several chromosomes in the *scid* cell line were associated with their p-arms to form the so called multi-branched chromosome configuration. The frequency of such multi-branched chromosomes in the *scid* cell line was about 10%.[24] A puzzling observation that longer telomeres and telomere fusions occur in the same cell line has prompted the speculation that probably the telomere telomerase complex may not be efficient in preventing end-to-end fusions in this particular cell line.[24] Telomere elongation was seen at both p- and q-arms of the chromosomes suggesting aberrant recombination or defective telomere capping function[24,25] (Hande, unpublished results).

To further investigate whether such a difference in telomere length also exists under in vivo condition, the telomere length was analysed in primary cells for different strains of *scid* mice. In all the strains of *scid* mice with their parental strains we have studied, 1.5 to 2 times longer telomeres could be detected.[25] In contrast to *scid* cell lines, bone marrow cells from *scid* mice did not exhibit telomere fusions (Hande, unpublished observation). It should also be noted that the difference in telomere length is wider in the cell lines (ten times) compared to the primary cells (1.5 to 2 times). Bailey et al[26] reported a similar occurrence of telomere fusions in a *scid* cell line though telomere length was not measured in that study. *scid* mice are deficient in the enzyme DNA-PK (DNA-dependent protein kinase) as a result of the mutation in the gene encoding the catalytic subunit (DNA-PKcs) of this enzyme. Our results on *scid* cell lines and primary cells from *scid* mice pointed to the possibility that DNA-PKcs either alone or in complex with other proteins in the nonhomologous end joining (NHEJ) pathway (see later) may, directly or indirectly, be involved in telomere length regulation in mammalian cells.

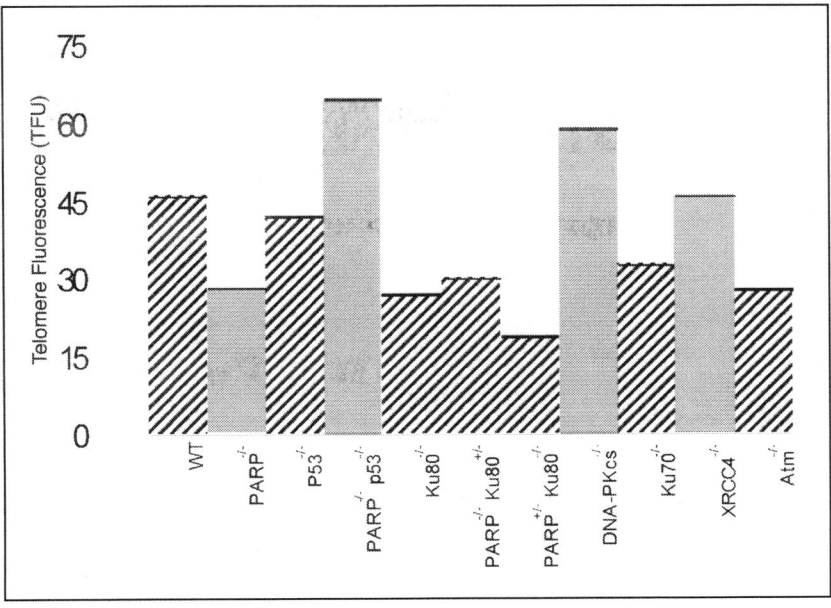

Figure 1. Telomere length dynamics in primary cells from DNA repair deficient mice. Reprinted from Hande[59] with permission; ©2004 S. Karger AG, Basel.

Nonhomologous End Joining (NHEJ) Complex and Telomeres

Eukaryotic cells use two different pathways for repairing DNA double strand breaks: homologous recombination (HR) and nonhomologous end joining (NHEJ). A key component of the homologous recombination process is Rad51,[27] a 50-kDa protein whose expression is cell-cycle dependent, peaking at the S/G_2 boundary.[28] The NHEJ pathway requires the activity of DNA-PK, a multimeric serine-threonine kinase composed of a catalytic subunit (DNA-PKcs) and two regulatory subunits (Ku70 and Ku80). Ku70 and Ku80 are able to recognise and bind to DNA DSB and then activate DNA-PKcs.[29] Another protein that plays a key role in NHEJ is XRCC4,[30] which interacts with, and probably controls the function of ligase IV.

DNA repair by NHEJ relies on the Ku70:Ku80 heterodimer in species ranging from yeast to man. In *Saccharomyces cerevisiae* and *Schizosaccharomyces pombe,* Ku also controls telomere functions. Based on our observation in *scid* mouse cells and on the studies in yeast, an analogous role for mammalian Ku complex in telomere maintenance can not be ruled out. Ku70, Ku80, and DNA-PKcs, with which Ku interacts, are associated in vivo with telomeric DNA in several human cell types and we have shown earlier that these associations are not significantly affected by DNA-damaging agents.[31] It was also demonstrated that inactivation of Ku80 or Ku70 in the mouse yields telomeric shortening in various primary cell types at different developmental stages (Fig. 1).[31] By contrast, telomere length is not altered in cells impaired in XRCC4 or DNA ligase IV, two other NHEJ components (Fig. 1). We also observe higher genomic instability in Ku-deficient cells than in XRCC4-null cells (Fig. 2). This study has suggested that chromosomal instability of Ku-deficient cells results from a combination of compromised telomere stability and defective NHEJ.[31]

Myung et al[32] have demonstrated that the functional inactivation of even a single allele of Ku86 in human somatic cells results in a profound telomere loss, which is accompanied by an increase in chromosomal fusions, translocations, and genomic instability measured by telomere

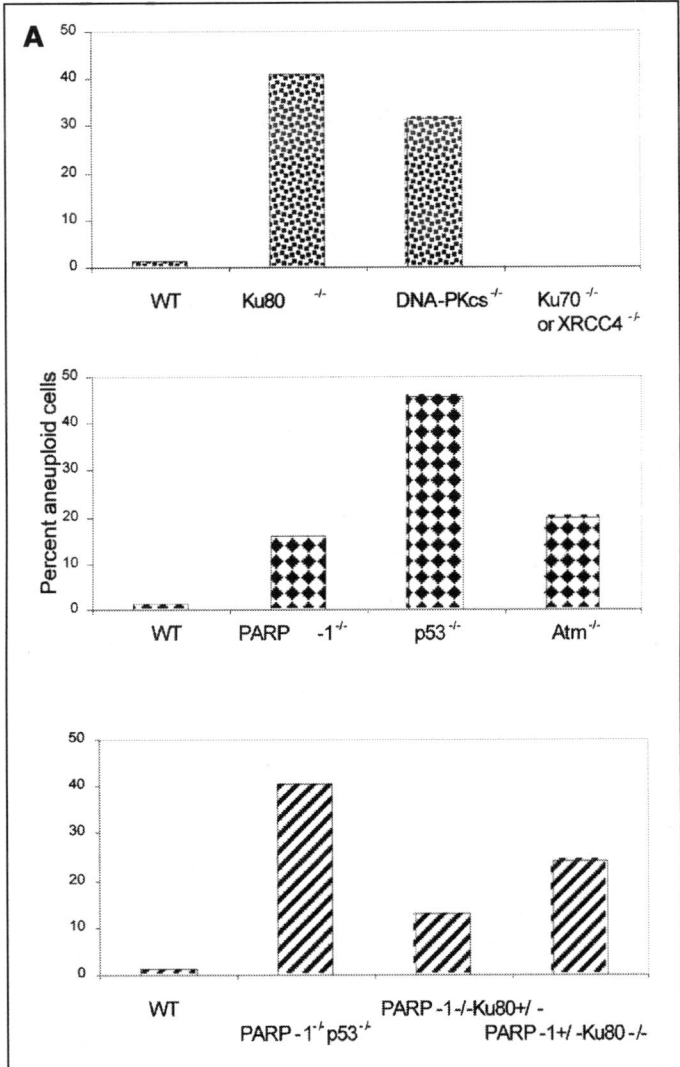

Figure 2. A) Percent aneuploidy in normal and DNA repair deficient mouse cells. Figure continued on next page.

FISH and spectral karyotyping. Taken together, the data available till now support the concept that Ku86, separate from its role in nonhomologous end joining, performs the additional function in human somatic cells of suppressing genomic instability through the regulation of telomere length.

The telomere repeat binding protein, TRF1 and Ku form a complex at the telomere sites.[33] The Ku and TRF1 complex determined by coimmunoprecipitation experiments is found to have a specific high-affinity interaction in human cells. Ku does not bind to telomeric DNA directly but localises to telomeric repeats via its interaction with TRF1. Primary mouse embryonic fibroblasts, deficient for Ku80, accumulated a large percentage of telomere fusions (Fig. 2),

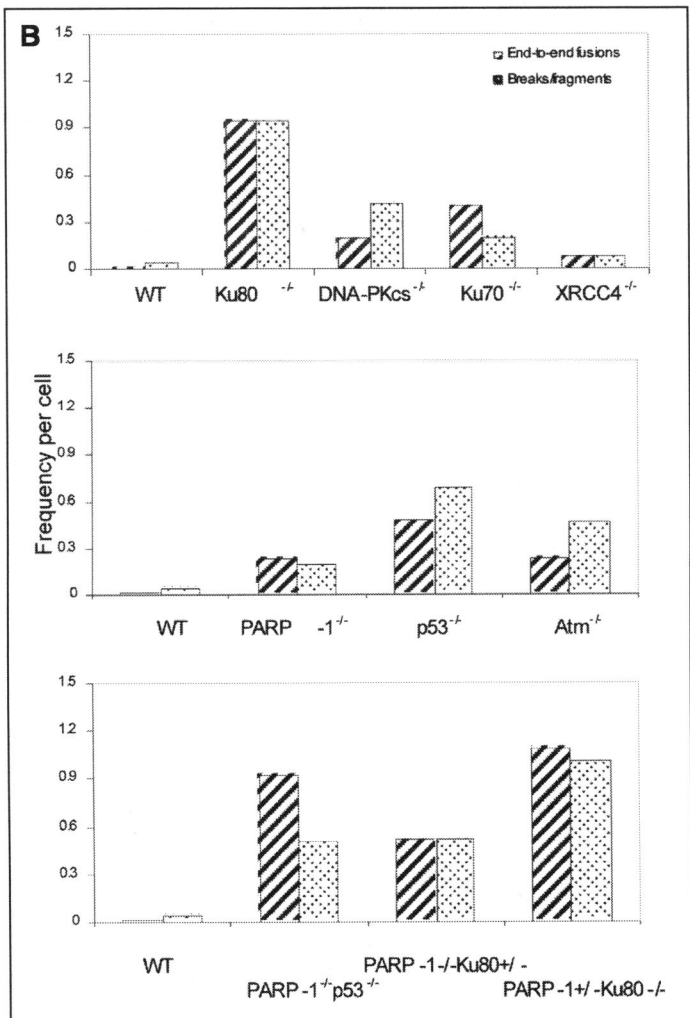

Figure 2, continued. B) Chromosome end-to-end fusions and fragments/breaks in normal and DNA repair deficient mouse cells. Spontaneous chromosome instability in primary embryonic fibroblasts from DNA repair-deficient mice. Aneuploidy could not be detected in mouse embryonic stem cells (XRCC4$^{-/-}$ and Ku70$^{-/-}$) as these cells to be intrinsically aneuploid with an average of 41 or 42 chromosomes per metaphase.

demonstrating that Ku plays a critical role in telomere capping in mammalian cells.[31,33] It was proposed that Ku localizes to internal regions of the telomere via a high-affinity interaction with TRF1.

The DNA-dependent protein kinase catalytic subunit (DNA-PKcs) is critical for DNA repair via the nonhomologous end joining pathway. As explained above, bone marrow cells and spontaneously transformed fibroblasts from *scid* mice have defects in telomere maintenance.[24,25] The genetically defective *scid* mouse arose spontaneously from its parental strain CB17. One known genomic alteration in *scid* mice is a truncation of the extreme carboxyl terminus of DNA-PKcs, but other as yet unidentified alterations may also exist. In another study, we used a defined system, the DNA-PKcs knockout mouse, to investigate specifically the role played by DNA-PKcs in telomere maintenance. Primary mouse embryonic fibroblasts (MEFs) and primary cultured kidney cells from 6-8 month-old DNA-PKcs-deficient mice accumulated a large number of telomere fusions, yet still retain wild-type telomere length (Figs. 1,2).[34] Thus, the phenotype of this defect separates the two-telomere related phenotypes, capping, and length maintenance. DNA-PKcs-deficient MEFs also exhibited elevated levels of chromosome fragments and breaks, which correlated with increased telomere fusions. Based on the high levels of telomere fusions observed in DNA-PKcs deficient cells (e.g., Fig. 3f), it was concluded that DNA-PKcs plays an important capping role at the mammalian telomere.[34] Recently, Espejel et al,[35] in a life-long follow-up study of DNA-PKcs-defective mice, observed that DNA-PKcs-defective mice had a shorter life span and showed an earlier onset of ageing-related pathologies than the corresponding wild-type littermates. In addition, DNA-PKcs ablation was associated with a markedly higher incidence of T-lymphomas and infections.[35] Therefore, these data link the dual role of DNA-PKcs in DNA repair and telomere length maintenance to organismal ageing and cancer.

PARP-1 and Telomeres

In most eukaryotes, poly(ADP-ribose) polymerase (PARP) recognises DNA strand interruptions generated in vivo. DNA binding by PARP triggers primarily its own modification by the sequential addition of ADP-ribose units to form polymers; this modification, in turn, causes the release of PARP from DNA ends.[36] Studies on the effects of the disruption of the gene encoding PARP-1 (*Adprt1*) in mice have demonstrated roles for PARP in recovery from DNA damage and in suppressing recombination processes involving DNA ends.[37-41]

Telomeres situated at the natural termini of chromosomes are therefore potential targets for PARP. Telomere shortening was seen in different genetic backgrounds and in different tissues, both from embryos and adult mice (Fig. 1)[42] without any change in the in vitro telomerase activity. Furthermore, cytogenetic analysis of mouse embryonic fibroblasts has revealed that lack of PARP-1 is associated with severe chromosomal instability, characterised by increased frequencies of chromosome fusions and aneuploidy (Figs. 2,3). The absence of PARP-1 did not affect the presence of single-strand overhangs, naturally present at the ends of telomeres. The above study has therefore revealed an unanticipated role for PARP-1 in telomere length regulation; though a later study using PARP-1 knockout mice with a different genetic background[43] did not reveal any significant change in telomere length. However, it cannot be ruled out that different genetic background mice and different mutations of the same gene might yield different results. Studies are presently in progress in my laboratory to evaluate the role of PARP-1 to address this issue. It will also be interesting to study the importance of other PARP molecules in genome stability.

Ataxia Telangiectasia Mutated (ATM) and Telomeres

Ataxia-telangiectasia (AT) is an autosomally recessive human genetic disease with pleiotropic defects such as neurological degeneration, immunodeficiency, chromosomal instability, cancer susceptibility and premature ageing. Cells derived from AT patients and ataxia-telangiectasia mutated (ATM)-deficient mice show slow growth in culture and premature senescence. ATM, which belongs to the PI3 (phosphatidylinositol 3) kinase family along with DNA-PK, plays a major role in signalling the p53 response to DNA strand breaks. Telomere maintenance is perturbed in yeast strains lacking genes homologous to ATM and cells from patients with AT have short telomeres.[16] We examined the length of individual telomeres in cells from *Atm*[-/-] mice by fluorescence in situ hybridisation techniques. Telomeres were extensively shortened in multiple tissues of *Atm*[-/-] mice[44] (Fig. 1). More than the expected number of

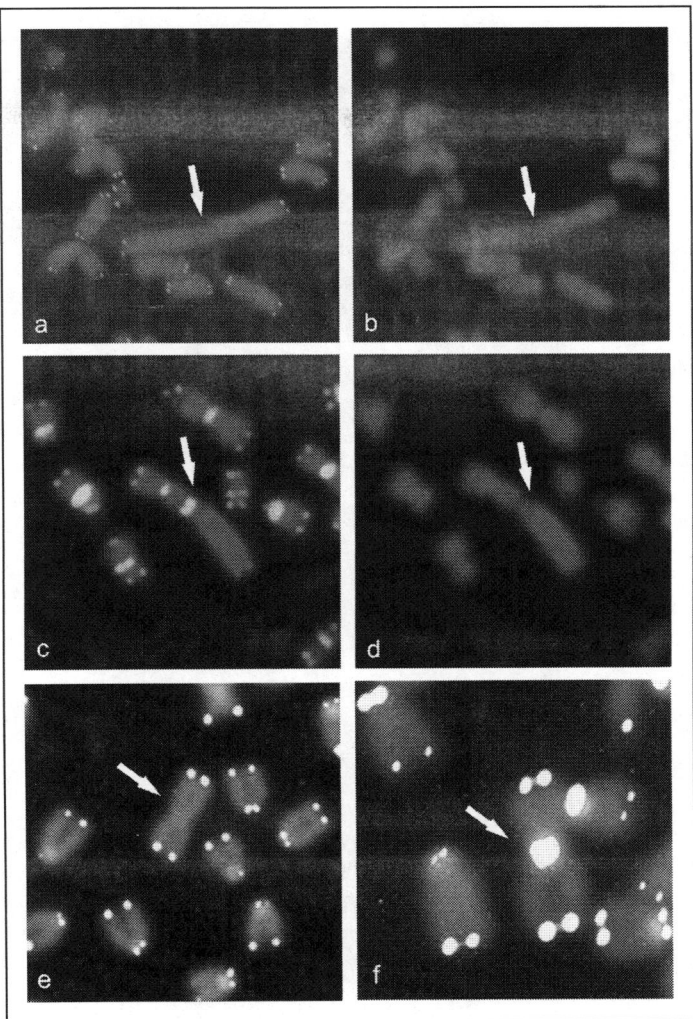

Figure 3. Telomere-mediated chromosome integrity in mammalian cells lacking telomerase or DNA repair factors. a) A typical end-to-end fusion detected in primary fibroblasts from Nijmegen Breakage Syndrome patients. Partial metaphase spread showing telomere signals on chromosomes after FISH with Cy3-labelled (CCCTAAA)$_3$ PNA probe and the arrow points to a fusion event. b) The same image as in (a) under DAPI filter. c) A dicentric (Dic) chromosome (pointed arrow) in a human metaphase spread showing telomere signals (red) at the termini and and two centromeres (green) along the chromosome arms (d) Image shown in (c) under DAPI. e) A typical Robertsonian fusion like configuration event due to the loss of telomeres at the p-arm of a mouse chromosome and fusion between the centromeres. No telomeres could be detected at the fusion point. f) An example of a Robertsonian fusion-like configuration (pointed arrow) with telomeres at the fusion point (RLCT) which could have been generated by the telomere dysfunction and the subsequent loss of capping function in these cells. These RLCs are generated either by complete loss of telomeres on p-arm of the chromosomes (telomere erosion) or fusion of different chromosome with telomeres at fusion point (capping function). e,f) Reproduced from Hande[59] with permission; ©2004 S. Karger AG, Basel. A color version of this figure is available online at www.Eurekah.com.

telomere signals was observed in interphase nuclei of *Atm*[-/-] mouse fibroblasts. Signals corresponding to 5-25 kb of telomeric DNA that were not associated with chromosomes were also noticed in *Atm*[-/-] metaphase spreads. Extrachromosomal telomeric DNA was also detected in fibroblasts from AT patients and may represent fragmented telomeres or by-products of defective replication of telomeric DNA. These results suggested a role for ATM in telomere maintenance and replication, which may contribute to the poor growth of *Atm*[-/-] cells and increased tumour incidence in both AT patients and *Atm*[-/-] mice.[44] Lustig[45] has suggested that formation of circular telomeric fragments could be due to the telomere rapid deletion as was seen in yeast. Recently, Herbig et al[46] looked at the association of ATM with the DNA damage foci induced by dysfunctional telomere (telomere induced foci) in senescent cells[47,48] and found that ATM colocalises with most senescent telomere induced foci.[46] Oxidative damage has been suggested to be one of the factors that influences the telomere status in AT cells. We are currently studying the precise role of chronic oxidative damage in normal and ATM[-/-] cells and in *Atm*[-/-] mice with special emphasis on genome stability and ageing.

Concerted Roles of PARP-1 and p53 on Telomeres

Genomic instability is often caused by mutations in genes that are involved in DNA repair and/or cell cycle checkpoints, and it plays an important role in tumourigenesis. To confirm our previous observation of telomere shortening in PARP-1 mice, we took advantage of PARP-1 and p53 double knockout mice to study the telomere-related chromosome instability and tumourigenesis in these mice. As PARP is thought to protect genomic stability,[38] its functional interaction with the guardian of the genome, p53 was tested using PARP-1[-/-]p53[-/-] mice. Compared to single-mutant cells, PARP-1 and p53 double-mutant cells exhibit many severe chromosome aberrations, including a high degree of aneuploidy, fragmentations, and end-to-end fusions, which to a large extent is attributable to telomere dysfunction.[49] While *PARP*[-/-] cells showed telomere shortening and *p53*[-/-] cells showed normal telomere length, inactivation of *PARP-1* in *p53*[-/-] cells surprisingly resulted in very long and heterogeneous telomeres, suggesting a functional interplay between PARP-1 and p53 at the telomeres (Fig. 1). Strikingly, PARP-1 deficiency widens the tumour spectrum in mice deficient in p53, resulting in a high frequency of carcinomas in the mammary gland, lung, prostate, and skin, as well as brain tumours. The enhanced tumourigenesis is likely to be caused by PARP-1 deficiency, which facilitates the loss of function of tumour suppressor genes as demonstrated by a high rate of loss of heterozygosity at the p53 locus in these tumours.[49] These results indicated that PARP-1 and p53 interact to maintain genome integrity and identify PARP as a cofactor for suppressing tumourigenesis. It is however not clear whether the long and heterogeneous telomeres are responsible for the wider tumour spectrum in these double knockout mice.

Interplay between PARP and Ku80 on Telomere-Chromosome Integrity

PARP-1 and Ku80 null cells showed telomere shortening associated with chromosome instability. Interestingly, haplo-insufficiency of PARP-1 in Ku80 null cells caused more severe telomere shortening and accumulation of chromosome abnormalities compared to either PARP-1 or Ku80 single null cells. Cytogenetic analysis of these cells revealed that many chromosome ends lack detectable telomeres as well. These results demonstrate that DNA break-sensing molecules, PARP-1 and Ku80, synergistically function at telomeres and play an important role in the maintenance of chromosome integrity. More importantly haplo-insufficiency of Ku80 in *PARP-1*[-/-] mice promoted the development of hepatocellular adenoma and hepatocellular carcinoma (HCC).[50] These tumours exhibited a multistage tumour progression associated with the loss of E-cadherin expression and the mutation of beta-catenin. Cytogenetic analysis revealed that Ku80 heterozygosity elevated chromosomal instability in *PARP-1*[-/-] cells and that these liver tumours harboured a high degree of chromosomal aberrations including fragmentations, end-to-end fusions, and recurrent nonreciprocal translocations. These features are reminiscent of human HCC. Taken together, these data implicate a synergistic function of Ku80 and PARP-1 in minimising chromosome aberrations and cancer development.

Homologous Recombination Proteins and Telomeres

The homologous recombination (HR) DNA repair pathway participates in telomere length maintenance in yeast but its putative role at mammalian telomeres is currently being studied. Strong evidence for telomere loss in the murine hypomorphic Rad50(S/S) homozygous mutant is shown by the frequent formation of fusions between chromosomes with short telomeres, which is consistent with a rapid loss of the originally elongated telomere.[51] Another component of homologous recombination pathway, NBS1 has been implicated in telomere maintenance. It has been shown that cells from NBS patients have strong telomere phenotypes. NBS cells have significantly shorter telomeres than their normal counterparts.[52] A rapid rate of telomere loss in cell lines from Nijmegen breakage syndrome (NBS) patients (who carry a nonnull NBS1 allele), might be caused by a combination of replicative loss, other mechanisms of attrition and an elevated rate of telomeric (end to end) fusions[53] (Hande et al unpublished results; (Fig. 3a,b). Introduction of both NBS1 and hTERT (telomerase catalytic subunit) genes was necessary to rescue the telomere shortening phenotypes in fibroblasts from NBS patients[52] indicating a strong interaction between NBS1 and telomerase complex. NBS1 is present at meiotic telomeres in mammalian cells.[54]

Jaco et al[55] have shown that Rad54-deficient mice show significantly shorter telomeres than wild-type controls indicating that Rad54 activity plays an essential role in telomere length maintenance in mammals.[55] Recently, Tarsounas et al[56] have reported that Rad51D is also involved in telomere maintenance. RAD51D was shown to localize to the telomeres of both meiotic and somatic cells. *Rad51d⁻/⁻ Trp53⁻/⁻* double knockout mouse embryonic fibroblasts exhibited telomere DNA repeat shortening compared to *Trp53⁻/⁻* or wild-type controls accompanied by elevated levels of chromosomal aberrations.

Concluding Remarks

It is evident from the telomere-chromosome data and associated genome instability data from DNA repair deficient mouse cells that the DNA damage/repair and signalling molecules play a vital role in the protection of telomeres and chromosomes and thereby maintain the genome integrity. As indicated in Figure 1, there seems to be a modest loss of telomeres in the cells lacking some of the DNA repair factors. This loss is approximately 30 to 40% of the original telomeres. Data on telomerase negative embryonic stem cells and embryonic fibroblasts lacking telomerase RNA indicated that a loss of approximately 60 to 70% of telomeric repeats was needed for the cells to become fusigenic and to induce the chromosome end-to-end fusion events in these cells. The telomeres are maintained by telomerase-independent mechanisms in *mTER⁻/⁻* mouse cells at later passages. Based on the above observations, it is tempting to speculate that at least in mice, a fraction of telomeres are also maintained by the factors involved in DNA repair/damage response and DNA damage signalling molecules. However, both telomerase-dependent and telomerase-independent mechanisms coexist to maintain telomeres in the mouse cells. It is plausible that telomere maintenance will be compromised by one pathway in the absence of another pathway. More interestingly, loss of DNA repair factors will render the cells to lose telomeres and a majority of these cells exhibit premature senescence. On the other hand, telomerase negative mouse cells escape senescence and could be immortalised in vitro through accumulation of chromosome instability. These two pathways interact with each other very efficiently and loss of either will lead to severe telomere mediated chromosome

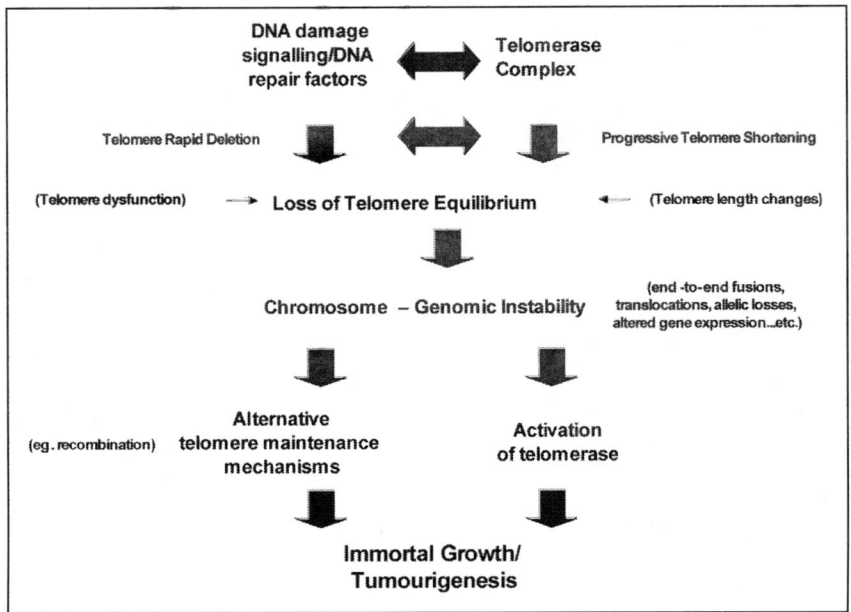

Figure 4. Role(s) of DNA damage signalling molecules and DNA repair factors in telomere mediated chromosome-genomic instability and tumourigenesis. Modified rom Hande[59] with permission; ©2004 S. Karger AG, Basel.

instability leading to tumourigenesis in mice (Fig. 4). Mice lacking genes involved in either DNA repair complex and/or telomerase complex develop spontaneous tumours and cells from these mice show telomere related chromosome abnormalities. Loss of telomere homeostasis might have contributed to the occurrence of severe chromosome instability in these mice, which led to the spontaneous development of tumours. However, given the many pathways in which DNA damage/repair and signalling molecules are involved, telomeric loss defects might represent a direct or indirect effect of the primary defect.

A study by d'Adda di Fagagna et al[47] documented the appearance of DNA damage-inducible foci in senescent primary fibroblasts and provided the functional evidence that these foci contribute directly to the senescent phenotype. This study has established the molecular and mechanistic similarities between critically shortened and uncapped telomeres and that the telomere attrition does induce a DNA damage response in primary human cells.[47] Components of the DNA damage response associated with DNA end-to-end joining or homologous recombination have been proposed to mediate alterations in telomere structure believed to be required for telomerase access.[57] These activities have also been linked to 'telomere rapid deletion' in yeast, which has been proposed to reset telomere size during meiosis.[45] Telomere rapid deletion has been defined as an end-mediated intrachromatid homologous-recombination event that results in a deleted telomere and a linear or circular by-product.[45] Catastrophic loss of telomere sequences in mouse cells lacking key DNA repair proteins, as observed in our studies with DNA repair deficient mouse cells, might reflect the above mentioned telomere rapid deletion process. However, this process needs to be thoroughly investigated in defined mammalian systems. Many of the DNA repair deficient mice or mouse cells (e.g., $Ku80^{-/-}$ and $Atm^{-/-}$) do exhibit premature ageing phenotype and cancer susceptibility.[58] Therefore, telomere rapid deletion and subsequent defects in genome maintenance systems in the repair deficient mice might contribute to the observed tumour susceptibility phenotype and accelerated ageing.

Acknowledgements

I would like to express my gratitude to Dr. Adayabalam S. Balajee, Centre for Radiological Research, Columbia University, New York, USA for his thoughtful suggestions and criticisms. Current research in my laboratory (Genome Stability Laboratory) is supported by grants from Academic Research Fund, National University of Singapore and Oncology Research Institute, Office of Life Sciences, National University Medical Institutes, Singapore. Dr. Anuradha Poonepalli and Mr. Aik Kia Khaw are acknowledged for their help during the manuscript preparation. Thanks are also due to Dr. Veena Hande for critically reading the manuscript.

References
1. McEachern MJ, Krauskopf A, Blackburn EH. Telomeres and their control. Annu Rev Genet 2000; 34:331-358.
2. Griffith JD, Comeau L, Rosenfield S et al. Mammalian telomeres end in a large duplex loop. Cell 1999; 97(4):503-14.
3. Moyzis RK, Buckingham JM, Cram LS et al. A highly conserved repetitive DNA sequence, (TTAGGG)n, present at the telomeres of human chromosomes. Proc Natl Acad Sci USA 1988; 85(18):6622-6.
4. Blackburn EH. Structure and function of telomeres. Nature 1991; 350(6319):569-73.
5. Greider CW. Telomere length regulation. Annu Rev Biochem 1996; 65:337-65.
6. Zakian VA. Telomeres: Beginning to understand the end. Science 1995; 270(5242):1601-7.
7. Greider CW, Blackburn EH. Identification of a specific telomere terminal transferase activity in Tetrahymena extracts. Cell 1985; 43(2 Pt 1):405-13.
8. Lingner J, Hughes TR, Shevchenko A et al. Reverse transcriptase motifs in the catalytic subunit of telomerase. Science 1997; 276(5312):561-7.
9. Blasco MA, Lee HW, Hande MP et al. Telomere shortening and tumor formation by mouse cells lacking telomerase RNA. Cell 1997; 91(1):25-34.
10. Heacock M, Spangler E, Riha K et al. Molecular analysis of telomere fusions in Arabidopsis: Multiple pathways for chromosome end-joining. EMBO J 2004; 23(11):2304-2313.
11. Greider CW. Telomeres. Curr Opin Cell Biol 1991; 3(3):444-51.
12. Allsopp RC, Vaziri H, Patterson C et al. Telomere length predicts replicative capacity of human fibroblasts. Proc Natl Acad Sci USA 1992; 89(21):10114-8.
13. Harley CB, Futcher AB, Greider CW. Telomeres shorten during ageing of human fibroblasts. Nature 1990; 345(6274):458-60.
14. Granger MP, Wright WE, Shay JW. Telomerase in cancer and aging. Crit Rev Oncol Hematol 2002; 41(1):29-40.
15. Kruk PA, Rampino NJ, Bohr VA. DNA damage and repair in telomeres: Relation to aging. Proc Natl Acad Sci USA 1995; 92(1):258-262.
16. Metcalfe JA, Parkhill J, Campbell L et al. Accelerated telomere shortening in ataxia telangiectasia. Nat Genet 1996; 13(3):350-3.
17. Schulz VP, Zakian VA, Ogburn CE et al. Accelerated loss of telomeric repeats may not explain accelerated replicative decline of Werner syndrome cells. Hum Genet 1996; 97(6):750-4.
18. Baird DM, Davis T, Rowson J et al. Normal telomere erosion rates at the single cell level in Werner syndrome fibroblast cells. Hum Mol Genet 2004.
19. de Lange T, DePinho RA. Unlimited mileage from telomerase? Science 1999; 283(5404):947-9.
20. Artandi SE, DePinho RA. A critical role for telomeres in suppressing and facilitating carcinogenesis. Curr Opin Genet Dev 2000; 10(1):39-46.
21. Lee HW, Blasco MA, Gottlieb GJ et al. Essential role of mouse telomerase in highly proliferative organs. Nature 1998; 392(6676):569-74.
22. Hande MP, Samper E, Lansdorp P et al. Telomere length dynamics and chromosomal instability in cells derived from telomerase null mice. J Cell Biol 1999; 144(4):589-601.

23. Niida H, Shinkai Y, Hande MP et al. Telomere maintenance in telomerase-deficient mouse embryonic stem cells: Characterization of an amplified telomeric DNA. Mol Cell Biol 2000; 20(11):4115-27.
24. Slijepcevic P, Hande MP, Bouffler SD et al. Telomere length, chromatin structure and chromosome fusigenic potential. Chromosoma 1997; 106(7):413-21.
25. Hande P, Slijepcevic P, Silver A et al. Elongated telomeres in scid mice. Genomics 1999; 56(2):221-3.
26. Bailey SM, Meyne J, Chen DJ et al. DNA double-strand break repair proteins are required to cap the ends of mammalian chromosomes. Proc Natl Acad Sci USA 1999; 96(26):14899-904.
27. Baumann P, Benson FE, West SC. Human Rad51 protein promotes ATP-dependent homologous pairing and strand transfer reactions in vitro. Cell 1996; 87(4):757-766.
28. Yamamoto A, Taki T, Yagi H et al. Cell cycle-dependent expression of the mouse *Rad51* gene in proliferating cells. Mol Gen Genet 1996; 251(1):1-12.
29. Critchlow SE, Jackson SP. DNA end-joining: From yeast to man. Trends Biochem Sci 1998; 23(10):394-8.
30. Critchlow SE, Bowater RP, Jackson SP. Mammalian DNA double-strand break repair protein XRCC4 interacts with DNA ligase IV. Curr Biol 1997; 7(8):588-98.
31. d'Adda di Fagagna F, Hande MP, Tong WM et al. Effects of DNA nonhomologous end-joining factors on telomere length and chromosomal stability in mammalian cells. Curr Biol 2001; 11(15):1192-6.
32. Myung K, Ghosh G, Fattah FJ et al. Regulation of telomere length and suppression of genomic instability in human somatic cells by Ku86. Mol Cell Biol 2004; 24(11):5050-5059.
33. Hsu HL, Gilley D, Galande SA et al. Ku acts in a unique way at the mammalian telomere to prevent end joining. Genes Dev 2000; 14(22):2807-12.
34. Gilley D, Tanaka H, Hande MP et al. DNA-PKcs is critical for telomere capping. Proc Natl Acad Sci USA 2001; 98(26):15084-8.
35. Espejel S, Martin M, Klatt P et al. Shorter telomeres, accelerated ageing and increased lymphoma in DNA-PKcs-deficient mice. EMBO Rep 2004; 5(5):503-509.
36. Lindahl T. Recognition and processing of damaged DNA. J Cell Sci Suppl 1995; 19:73-7.
37. de Murcia JM, Niedergang C, Trucco C et al. Requirement of poly(ADP-ribose) polymerase in recovery from DNA damage in mice and in cells. Proc Natl Acad Sci USA 1997; 94(14):7303-7.
38. Jeggo PA. DNA repair: PARP - another guardian angel? Curr Biol 1998; 8(2):49-51.
39. Lindahl T, Satoh MS, Dianov G. Enzymes acting at strand interruptions in DNA. Philos Trans R Soc Lond B Biol Sci 1995; 347(1319):57-62.
40. Morrison C, Smith GC, Stingl L et al. Genetic interaction between PARP and DNA-PK in V(D)J recombination and tumorigenesis. Nat Genet 1997; 17(4):479-82.
41. Wang ZQ, Stingl L, Morrison C et al. PARP is important for genomic stability but dispensable in apoptosis. Genes Dev 1997; 11(18):2347-58.
42. d'Adda di Fagagna F, Hande MP, Tong WM et al. Functions of poly(ADP-ribose) polymerase in controlling telomere length and chromosomal stability. Nat Genet 1999; 23(1):76-80.
43. Samper E, Goytisolo FA, Menissier-de Murcia J et al. Normal telomere length and chromosomal end capping in poly(ADP-ribose) polymerase-deficient mice and primary cells despite increased chromosomal instability. J Cell Biol 2001; 154(1):49-60.
44. Hande MP, Balajee AS, Tchirkov A et al. Extra-chromosomal telomeric DNA in cells from Atm($^{-/-}$) mice and patients with ataxia-telangiectasia. Hum Mol Genet 2001; 10(5):519-28.
45. Lustig AJ. Clues to catastrophic telomere loss in mammals from yeast telomere rapid deletion. Nat Rev Genet 2003; 4(11):916-923.
46. Herbig U, Jobling WA, Chen BP et al. Telomere shortening triggers senescence of human cells through a pathway involving ATM, p53, and p21(CIP1), but not p16(INK4a). Mol Cell 2004; 14(4):501-513.
47. d'Adda dF, Reaper PM, Clay-Farrace L et al. A DNA damage checkpoint response in telomere initiated senescence. Nature 2003; 426(6963):194-198.
48. Takai H, Smogorzewska A, de Lange T. DNA damage foci at dysfunctional telomeres. Curr Biol 2003; 13(17):1549-1556.

49. Tong WM, Hande MP, Lansdorp PM et al. DNA strand break-sensing molecule poly(ADP-Ribose) polymerase cooperates with p53 in telomere function, chromosome stability, and tumor suppression. Mol Cell Biol 2001; 21(12):4046-54.

50. Tong WM, Cortes U, Hande MP et al. Synergistic role of Ku80 and poly(ADP-ribose) polymerase in suppressing chromosomal aberrations and liver cancer formation. Cancer Res 2002; 62(23):6990-6.

51. Bender CF, Sikes ML, Sullivan R et al. Cancer predisposition and hematopoietic failure in Rad50(S/S) mice. Genes Dev 2002; 16(17):2237-2251.

52. Ranganathan V, Heine WF, Ciccone DN et al. Rescue of a telomere length defect of Nijmegen breakage syndrome cells requires NBS and telomerase catalytic subunit. Curr Biol 2001; 11(12):962-6.

53. Tauchi H, Matsuura S, Kobayashi J et al. Nijmegen breakage syndrome gene, *NBS1*, and molecular links to factors for genome stability. Oncogene 2002; 21(58):8967-8980.

54. Lombard DB, Guarente L. Nijmegen breakage syndrome disease protein and MRE11 at PML nuclear bodies and meiotic telomeres. Cancer Res 2000; 60(9):2331-4.

55. Jaco I, Munoz P, Goytisolo F et al. Role of mammalian Rad54 in telomere length maintenance. Mol Cell Biol 2003; 23(16):5572-80.

56. Tarsounas M, Munoz P, Claas A et al. Telomere maintenance requires the RAD51D recombination/repair protein. Cell 2004; 117(3):337-347.

57. Blackburn EH. Telomere states and cell fates. Nature 2000; 408(6808):53-56.

58. Hasty P, Campisi J, Hoeijmakers J et al. Aging and genome maintenance: Lessons from the mouse? Science 2003; 299(5611):1355-1359.

59. Hande MP. DNA repair factors and telomere-chromosome integrity in mammalian cells. Cytogenet Genome Res 2004; 104(1-4):116-122.

CHAPTER 10

Defective Solar Protection in Xeroderma Pigmentosum and Cockayne Syndrome Patients

Colette apRhys and Daniel Judge

Clinical Presentation

Xeroderma pigmentosum (XP),[1] Cockayne syndrome (CS)[2,3] and xeroderma pigmentosum-Cockayne syndrome (XP-CS)[4,5] are rare disorders with autosomal recessive inheritance, characterized by extreme sensitivity to sunlight. This sensitivity reflects the inadequate removal and/or repair of UV-induced lesions in the affected individual's DNA by nucleotide excision repair (NER).[6] While XP and CS share a defect in a common pathway, they are clinically distinct, albeit with some overlap resulting in the syndrome known as XP-CS.[7,8]

XP usually presents itself as severe sunburn between the child's first and second year.[9] Early freckling and premature aging of sun exposed skin is characteristic of XP, hence it's name meaning dry pigmented skin.[6] Affected individuals have a greater than 1,000 fold elevated risk of skin cancer, eg. basal cell carcinomas and malignant melanomas.[10] They also are at increased risk for damage to sun exposed parts of the eye such as the lids, conjunctiva, and cornea.[11] The lids become hyperpigmented and loose their lashes. Severe keratitis leads to corneal scarring and ulceration. Squamous cell carcinomas, basal cell carcinomas, and melanomas may occur at the ocular surface. Approximately 20% of XP patients show progressive neurological decline.[12] The abnormalities may be mild such as diminished deep tendon stretch reflexes, or severe such as microcephaly, mental retardation, spasticity, ataxia, seizures and progressive hearing loss. Postmortem analysis of these individuals shows loss of neurons in the cerebrum and cerebellum as well as in the spinal cord and peripheral nervous system.[8] In a single individual, some of the abnormal skin conditions of the XP phenotype are associated with the clinical features of trichothiodystrophy (TTD).[13]

CS is classified as a premature aging syndrome.[14] Affected individuals suffer from severe dwarfism with a stooped posture, severe cachexia, and mental retardation. They have thinning of hair and skin, large ears and nose, sunken eyes, cataracts, and dental caries giving them a wizened appearance.[2,15,16] Demyelination of neural axons in the brain and calcification of the brain is apparent postmortem.[17] Unlike XP, CS is not associated with an increased risk of cancer.[18] There are two forms[19,20] differing in the severity of the disease.

DNA Repair and Human Disease, edited by Adayabalam S. Balajee. ©2006 Landes Bioscience and Springer Science+Business Media.

Affected individuals with the "classic" late onset form, CS group A, appear normal at birth but acquire severe growth failure and developmental abnormalities within the first two years. These include progressive loss of hearing, vision, and central and peripheral nervous system function, resulting in severe disability. They die within the first 20 years.[21] Individuals with CS group B, suffer from prenatal growth failure with little or no postnatal neurological development.[17,22,23] They have congenital cataracts and other defects of the eye. They have early contractures of the spine and scoliosis. Their symptoms progressively worsen and they die within the first 7 years.[21]

XP-CS displays features of both XP and CS.[8,24] To date only a dozen cases have been identified. Affected individuals have the skin freckling and increased cancer susceptibility of XP, and also the neurological and developmental problems associated with CS.[25,26]

Human DNA Repair

A method examining unscheduled DNA synthesis (UDS) provided the first evidence for DNA repair in mammalian cells.[27] Cells in tissue culture were pulse labeled with [^{3}H]-thymidine, fixed and overlaid with photographic emulsion, then subjected to autoradiography. Only cells in S phase undergoing DNA replication incorporated label (scheduled DNA synthesis). Following UV irradiation, all the cells incorporated label but at a lower level (unscheduled DNA synthesis). Retrospectively, this observation reflected the excision of damaged DNA and DNA resynthesis. Jim Cleaver proved the first link between a human disease syndrome and DNA repair. He showed that the majority of cells from XP affected patients were deficient in UDS after UV irradiation.[6,28] By fusing cells from different patients and examining each pair for restoration of normal UDS after UV irradiation, he was able to assign the majority of patients to 7 "classical XP" complementation groups (A-G). The cells of the remaining patients appeared to have normal UDS and were categorized as a variant form of the disease, XP-V[29] (Table 1). These observations provided the foundation for elucidating the process of mammalian NER.

Table 1. XP and CS complementation groups

HCG	RCG	LOCUS	GENE	REPAIR STATUS
XPA		9q22	XPA	GGRd; TCRd
XPB	ERCC3	2q21	XPB	NERd
XPC		3p25	XPC	GGRd; TCRn
XPD	ERCC2	19q13	XPD	NERd
XPE		11p12-p11	DDB2	GGRd; TCRn
XPF	ERCC4	19q13	XPF	NERd
XPG	ERCC5	13q32	XPG	NERd
XPV		16p12-21	POLH	GGRn; TCRn; TLSd
CSA	ERCC8	5q12.1	CKN1	GGRn; TCRd
CSB	ERCC6	10q11	CSB	GGRn; TCRd

Human complementation group (HCG); Rodent complementation group (RCG); Global genomic repair deficient or normal (GGRd or n); Transcription coupled repair deficient or normal (TCRd or n); Nucleotide excision repair (GGR + TCR) deficient or normal (NERd or n); Translesion synthesis deficient or normal (TLSd).

Genes

Early work focused on isolating the genes involved in NER. Chinese hamster ovary cells and murine cells were mutagenized by DNA damaging agents, such as ethyl methanesulfonate (EMS) or UV irradiation, and screened for increased sensitivity to these agents.[30] In this manner, cell lines representing 11 rodent complementation groups (rcg1-rcg11) were isolated.[31] Fragments of human genomic DNA were stably integrated into these various cell lines. The transfected cells were selected for functional human genomic DNA complementation of NER by survival after UV treatment. In this manner, 6 excision-repair-cross-complementing genes (ERCC) were rescued, ERCC1-6.[32-37] ERCC2, ERRC3, and ERCC5 have proven to be the genes defective in XPD,[38] XPB[34] and XPG[39] respectively. ERCC6/CSB has proven to be a gene defective in a subset of patients affected with CS.[37] Transfection of a human expression cDNA library into affected human cells and screening for functional complementation identified the XPC gene[40] and another gene defective in a subset of CS patients, CSA.[41] The XPA gene was also identified by functional complementation but in this case total genomic DNA was transfected into human XPA cells.[42] Isolation of the gene responsible for the defect in XPE patients proved to be more difficult. Initially, a protein complex that binds UV-irradiated DNA was found to be lacking in XPE cells.[43] This protein complex, DDB, was later recognized as being composed of two subunits, DDB1 and DDB2. The cDNAs encoding both of these proteins were isolated.[44] Subsequently, it was shown that it is the DDB2 subunit that is mutated in XPE patients.[45]

Mutations Underlying the Phenotype

XPA-G,[44,46-51] XPVariant,[52] CSA[41] and CSB[37] have been chromosomally mapped and sequenced (Table 1). It is often difficult to predict the patients's clinical outcome based simply on his underlying genotype. Mutations in the same gene have been shown to have widely different clinical outcomes. For instance, defects in XPD can result in XP, XP/CS, or TTD phenotypes.[53,54] This allelic heterogeneity is further complicated by the fact that many patients are not homozygous for a mutation, but are compound heterozygotes. Additionally, mutations in different genes can underlie similar phenotypes. For instance, the CS phenotype has been shown to be caused not only by mutations in the CSA and CSB gene but also by mutations in the XPB, XPD, XPG genes.[8,21,26,55-57] These mutations include missense mutations, premature termination codons (PTCs), mutations at splice donor/acceptor sites, or deletions that usually shift the reading frame and generate incorrect amino acids substitutions and PTCs. An aberrant mRNA containing a PTC would likely be degraded by the Nonsense mRNA Decay (NMD) pathway resulting in little or no protein product and could, in effect, act as a null mutation. Differing efficiencies of NMD between individuals, and even within tissue and cell types, may confuse the issue by affecting the level of stability of mRNA encoding a truncated protein. An understanding of the precise mutation in the affected gene, and how it affects its role in NER and other cellular processes may be helpful as a clinical predictor.

The Cellular and Biochemical Phenotype of NER

Exposure to UV is acquired by incident sunlight and accidental irradiation by germicidal lamps. The UV spectrum ranges from 200-400 nm. This spectrum is divided into three sections, UVA (320-400 nm), UVB (280-320 nm) and UVC (200-280 nm). Incident solar radiation is comprised of UVA and UVB since UVC radiation cannot penetrate the atmosphere's ozone layer.[58] DNA absorbs UV in the range of 245-290 nm with a peak absorbance of 260 nm.[59] Therefore, UVB irradiation is the most significant environmental exposure resulting in DNA damage.[60] There is a typical "UV mutagenesis signature" first described for the p53 tumor suppressor gene mutated in 14% of skin tumors from normal

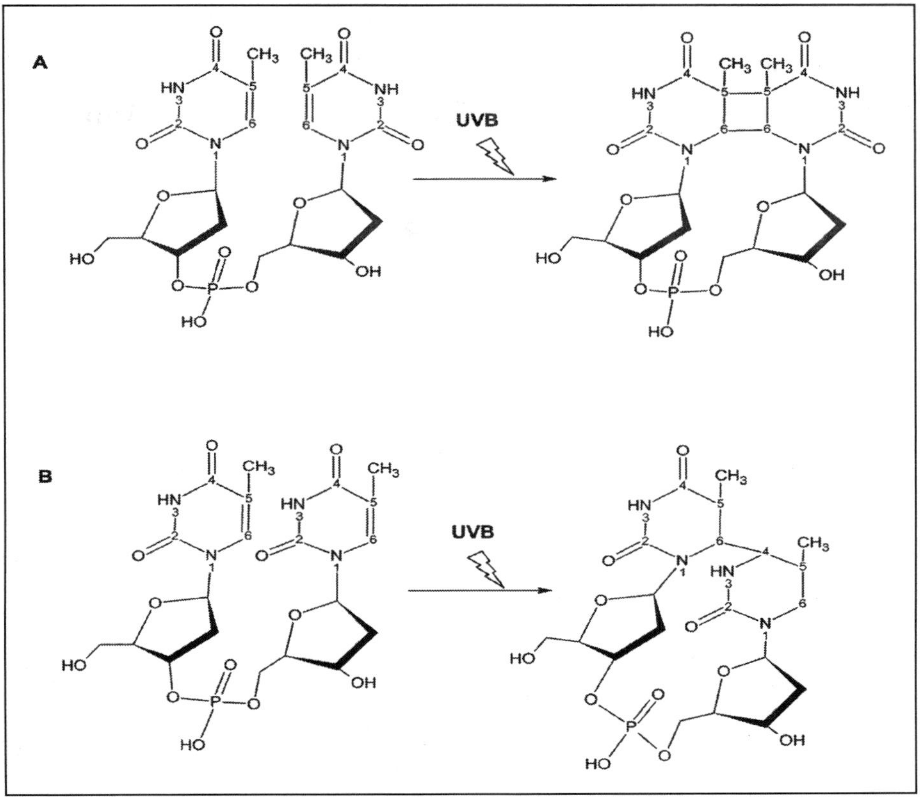

Figure 1. Structure of the major mutagenic lesions incurred by UVB or UVC. Absorption of UV light in the range of 200-320 nm induces mutagenic photoproducts between adjacent pyrimidines. A) Two adjacent pyrimidines can be converted to a cyclobutane dimer when a covalent bond forms between their C-4 and C-6 atoms resulting in a 4-member ring structure. B) [6-4] pyrimidine-pyrimidone photoproducts occur when a single covalent bond forms between the C-6 atom of one pyrimidine and the C-4 atom of a neighboring 3' pyrimidine.

individuals.[61,62] Mutations mainly occurred in dipyrimidine "hot spots" resulting in C to T transitions, 14% of which are CC to TT tandem mutations[62-64] and tandem mutations appear only at a frequency of 0.8% in[65] internal malignancies from normal individuals. In over 90% of skin tumors from "classical XP" affected patients deficient in UV damage repair, all mutations are at dipyrimidine sites and over 50% are CC to TT transitions.[66,67]

The major mutagenic DNA lesions incurred by UVB or UVC radiation are cyclobutane pyrimidine dimers (CPDs) and [6-4] pyrimidine-pyrimidone photoproducts (6-4PPs) (Fig. 1).[68] CPDs are generated from two adjacent pyrimidines when a covalent bond forms between each of their C-4 atoms and each of their C-6 atoms resulting in a 4 member ring structure that draws the bases closer together. The resulting distortion in the DNA causes a bend of about 30°.[69] T-T CPDs are most common followed by T-C or C-T dimers.[70] These pyrimidine dimers are found primarily in the *cis-syn* form although the *trans-syn* form can occur in single-stranded regions. 6-4 PP's occur when a single covalent bond forms between the C-6 atom of one pyrimidine and the C-4 position of a neighboring 3' pyrimidine.[71,72] This C-6 to C-4 linkage seriously distorts the DNA, bending it about 44°.[73] The 6-4 products of T-C are the most common followed by C-C and then T-T(59). These two adducts that distort DNA, block DNA replication by DNA polymerase and block transcriptional elongation by RNA polymerase. Failure to remove these lesions results in cessation of transcription,[74] cell cycle arrest,[75] genomic instability, or apoptosis.[76] In human cells, the mechanism of NER is responsible for the repair of these photoproducts. In cells from XP patients with the exception of cells from the variant group, removal of these lesions is impaired up to 90% of a normal rate.[77] As it turns out, more than 25 polypeptides are involved in NER, many of which have additional cellular functions.[78]

There are two converging pathways of NER. One transcription coupled repair (TCR) is responsible for the rapid repair of CPDs in the template strand of DNA undergoing transcription.[79-81] The other, global genomic repair (GGR) is responsible for general repair of the genome,[82] in particular 6-4 pyrimidine dimers. It is currently accepted that there is a sequential assembly of NER factors for both pathways,[82-84] differing only in the initial steps of damage recognition[85] (Fig. 2).

Damage Recognition

Recognition of damage by TCR relies on a RNA polymerase II molecule stalling at the site of damage.[76-86] TFIIH, a protein complex with 9 subunits including XPB and XPD, is required for unwinding DNA around the start site of transcription and maintaining 11 -13 base pairs in an open complex configuration during transcription initiation.[87] As RNA polymerase II begins to transverse along DNA, the open complex formation expands to 17 base pairs (transcription bubble) independent of TFIIH.[88] Upon encountering damage in the DNA template strand, the RNA pol II elongation complex arrests, sterically hindering the access of NER proteins. At this point, RNA pol II must be either displaced or removed from its template. Mutations in CSA or CSB result in deficient TCR but not GGR,[89-90] making their protein products likely candidates for "template clearance". The CSB protein, a SWI/SNF-like DNA-dependent ATPase has been shown to remodel chromatin and promote the addition of an extra nucleotide by RNA pol II.[91] It is plausible that CSB could modify the RNA pol II – DNA interface or push RNA pol II past the transcription blocking lesion or even remove RNA pol II from its DNA template. There is also evidence that RNA pol II stalled at a lesion or at a pause site is targeted for degradation by ubiquitination and that mutations in CSB inhibit this ubiquitination.[92] Additionally, the CSB protein may be involved in recruiting NER proteins to the damage site as it has been found in a cellular complex containing damaged DNA, RNA pol II, and TFIIH.[93,94] The CSB protein is required for translocation of the CSA protein to the nuclear matrix following UV irradiation.[95] The CSA protein contains 5 WD repeats suggesting that it is involved in protein - protein interactions.[96] It is found in a complex with XAB 2 (XPA binding-2) protein and the XPA protein.[97] It has been shown to interact with CSB and RNA pol II.[97] Its function maybe simply to act as a scaffold linking the transcriptional apparatus with the NER apparatus. A deficiency in TCR results in lesions left in nontranscribed genomic regions and in the nontemplate strand of transcribed genes.[89,90]

DNA damage recognition by GGR is not well defined possibly because the cell has multiple ways to demarcate DNA damage in nontranscribed strands of DNA. Two of the XP syndromes, XPE and XPC, are defective in GGR but not TCR[98-101] suggesting they are involved in the first step of damage recognition. Patients with XPE display mild sun sensitivity but are at high risk for skin cancer. Cells from these individuals show increased survival /decreased apoptosis but only a 50-80 % reduction in UDS after UV treatment.[102-103] XPE cells are predominately defective in CPD removal from nontranscribed DNA.[104] The 6-4PPs occur outside of chromatinized regions of DNA but CPDs can form within nucleosomal cores and within tightly packed chromatin making them particularly difficult to recognize and remove. Apparently, XPE cells lack a UV-damaged DNA binding activity (UV-DDB),[43] a heterodimeric protein

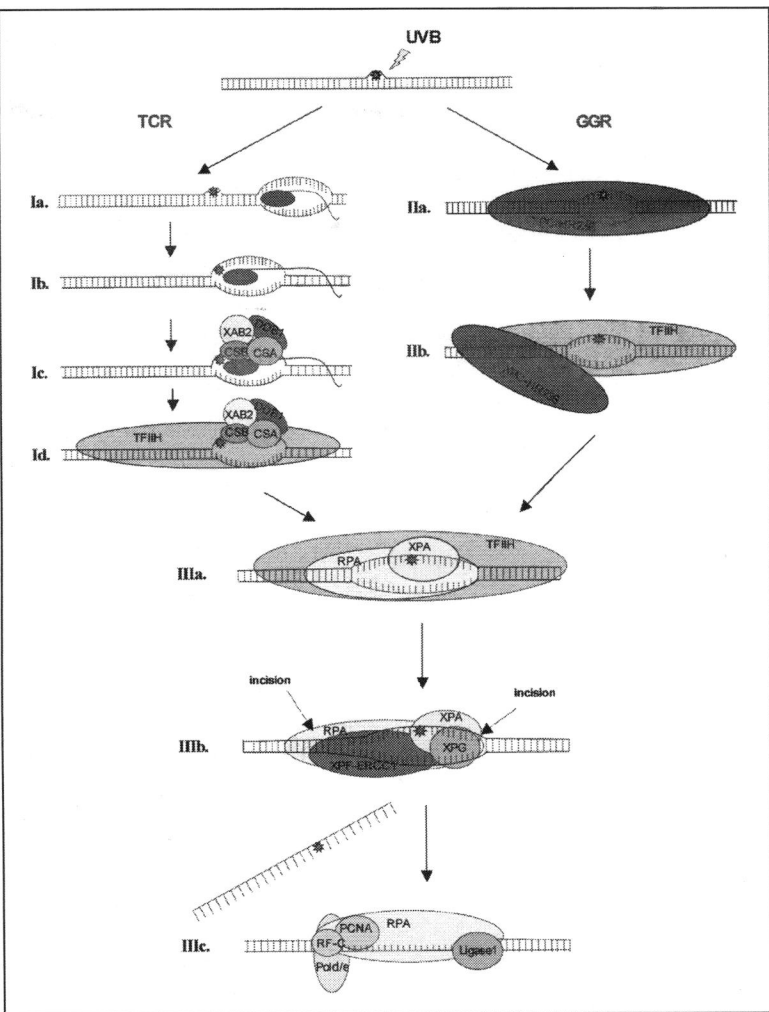

Figure 2. Model of human nucleotide excision repair (NER). Solar radiation in the UVB range damages DNA. There are two converging pathways of NER differing only in the initial steps of damage recognition. Global genomic repair (GGR) removes damage genome-wide. Transcription coupled repair (TCR) removes lesions in the template strand of transcribing genes. Ia) During TCR, RNA polymerase progresses along DNA in an open complex formation. Ib) Upon encountering damage in its template strand, RNA polymerase stalls. Ic) A multiprotein complex containing CSA, CSB, XAB2 and DDB1, forms at the lesion site and proceeds to clear the template strand of RNA polymerase. Id) CSB interacts with TFIIH in preparation for entering the common NER pathway. IIa) During GGR, the XPC-HR23B protein complex recognizes and binds to the damaged DNA. IIb) Binding of XPC-HR23B recruits TFIIH which, in turn, displaces XPC-HR23B in the 5' direction. The damaged substrate now can enter the repair pathway shared with TCR. IIIa) XPA together with RPA are recruited to the lesion either by the presence of XPC or CSB. In an ATP dependent manner, TFIIH further unwinds the DNA. The opening of the DNA is stabilized by XPA. IIIb) XPG joins the complex followed by XPF/ERCC1. Together, they incise 5' and 3' of the damage site. Subsequently, a 24-32 base oligonucleotide is excised and released. IIIc) RPA remains associated with the gapped DNA. RF-C loads PCNA onto the DNA and recruits a DNA polymerase to resynthesize the DNA. After the gap is filled, DNA Ligase1 ligates the newly synthesized DNA to the 3' flanking DNA.

composed of two subunits, DDB1 (p127) and DDB2 (p48).[103-105] The XPE gene encodes the smaller of the two proteins, the DDB2 subunit.[107] The amino acid sequence of DDB2 contains a region of WD40 repeats similar to those found in CSA.[108] Interestingly, DDB1 is found in two nearly identical complexes made up of about twelve polypeptides.[109] The only difference between them appears to be the presence of either CSA or DDB2. Microinjection of purified DDB1 containing complex into XPE cells restores UDS to normal levels, reflecting a role in GGR.[102,103,109] Microinjection of purified CSA containing complex into CS group A cells restores post-UV RNA synthesis recovery reflecting a role in TCR.[109] Purified UV-DDB binds with very high affinity to DNA substrates containing CPDs or 6-4 pps.[106] A CPD causes significantly less helical distortion in the DNA than does a 6-4 PP. DDB binding to a CPD increases the distortion, bending the DNA about 55°.[110] Surprisingly, its addition to in vitro excision assays has resulted in little to no increase in the efficiency of CPD or 6-4 PP removal.[111,112] DDB has been shown to bind CPD and 6-4PP in vivo[113,114] and to be required for XPC binding to CPD but not 6-4PP.[114] It has been proposed that DDB may act by recognizing DNA damage in vivo and remodel the local chromatin structure facilitating the access of XPC.[114,115]

In XPC cells, CPDs and 6-4PPs are removed only from the transcribed strand of active genes.[98,116] Adducts remain in transcriptionally inactive regions and in the nontranscribed strand of DNA. In normal cells, the XPC protein is found complexed with the HR23B protein[117] and centrin 2.[118] The purified heterodimer XPC-HR23B has been shown to bind various forms of DNA damage.[85,119-121] It has a greater affinity for lesions that induce a higher degree of helix distortion.[122] The complex recognizes a DNA structure containing double and single strand junctions with a strong preference for the ends of double stranded DNA containing a 3'OH.[123] When compared to an undamaged substrate, XPC-HR23B shows preferential binding to substrates containing 6-4PPs but not to CPDs.[120,124] However in a whole cell extract excision assay, repair of both 6-4PPs and CPDs is dependent on the presence of XPC.[120] The binding of XPC-HR23B to a lesion is probably the primary damage recognition step but there are other auxiliary factors such as DDB that act by making a lesion more accessible or recognizable by binding and distorting the DNA.

Open Complex Formation

Recently, the binding of XPC to the damage site has been shown to initiate opening of the DNA helix and to recruit TFIIH.[125] In the presence of ATP, the XPB and XPD helicase subunits of TFIIH further unwind the DNA.[125] During this process, the XPC-HR23B complex is displaced from the 3' side of the lesion allowing entry of XPA, a zinc finger DNA binding protein, and RPA to the site.[125] In GGR, both XPA and RPA are recruited to a lesion by the presence of XPC. TFIIH unwinds the DNA resulting in an open complex formation. The XPA protein interacts with TFIIH stabilizing the open complex. At the lesion site, RPA binds the undamaged single stranded DNA and XPA binds the damaged strand. The orientation of the XPA-RPA complex then guides the incision of the correct, damaged DNA strand.[126] It is an essential factor for both the TCR and GGR pathway as in its absence NER is abolished.[127] The demonstration of binding of the XPA protein to the CSB protein and to the p34 subunit of TFIIH provides an obvious, but not proven, connection to the TCR pathway.

Incision

In vitro studies have shown that XPA-RPA complex binds the core NER single stranded endonucleases, XPG[128] and ERCC1/XPF.[50,84,129] The XPG protein belongs to the flap endonuclease 1 (FEN1) family of structure specific nucleases that bind to single-stranded/double-stranded DNA junctions.[130] While other members of the family, Fen1 and exonuclease

1 (Exo1), can only bind DNA substrates with a free 5' single-stranded DNA end,[131,132] XPG can also bind bubble-like structures.[133] Its presence is required for fully open complex formation (approximately 30 nucleotides wide) around the lesion.[125,134] It's recruitment to a lesion is not dependent on the presence of XPA[83] but XPA-RPA is required for XPG to incise at the 5th-9th phosphodiester bond 3' of the lesion.[84] The arrival of XPG to the lesion site coincides with the departure of XPC-HR23B, which is able to engage in another round of NER.[84] XPG also binds to proliferating cell nuclear antigen (PCNA),[130,135] an essential factor in the resynthesis step of NER but not the incision step.[84] XPF forms a tight complex with ERCC1 in vivo.[136] XPF-ERCC1 cleaves Y-type DNA structures at the point where single stranded DNA branches off from double stranded DNA in the 5' to 3' direction.[137] Its recruitment is dependent on XPA.[83,138] This complex is responsible for incision at the 16th-24th phosphodiester bond 5' of the lesion.[139] The formation of a stable incision complex triggers the dual incision step and the subsequent release of XPF-ERCC1, XPG, TFIIH[84] and a 24-32 base oligonucleotide,[140] the length of which is determined by the type of damage.[139] RPA remains associated with the gapped DNA possibly protecting the ssDNA ends flanking the gap.[84]

DNA Resynthesis

PCNA is required for resynthesis of the gap generated by NER.[141] PCNA is made up of three monomer subunits arranged in a head to tail fashion forming a ring structure termed a clamp.[142,143] The inside of the ring is lined with basic residues that interact with the phosphate backbone of DNA allowing it to slide along the DNA.[142] Amino acid residues on the outer surface of the ring interact with DNA polymerase δ or ε preventing them from disassociating from the DNA.[144,145] Thus, PCNA enhances the processivity of these polymerases.[146] Replication Factor C (RF-C) is required for opening and closing the PCNA clamp around the DNA in an ATP dependent manner.[147,148] After loading PCNA onto the DNA template, RF-C disassociates from PCNA.[146] PCNA is loaded at the XPG incision site[135] and recruits pol δ or pol ε to resynthesize the excised DNA.[149,150] DNA synthesis proceeds until the end of the gap leaving a nick in the DNA at the 3' end. PCNA then recruits DNA Ligase I (Lig1) to perform the final step, ligating the newly synthesized DNA to the 3' flanking DNA.[151] Interestingly, a patient with a deficiency in Lig1 showed increased sensitivity to UV supporting the premise that it is Lig1 that is required for NER in vivo.[152]

Translesion Synthesis

In contrast to patients suffering from "classical XP", patients suffering from XP-Variant have normal levels of NER as determined by UDS. They have mild sun sensitivity but are still highly cancer prone.[153] Their symptoms present later in life, generally after 30 years old. After UV irradiation, fibroblasts from these patients show a deficit in post-replication repair (PRR),[154,155] loosely described as the appearance of low molecular weight DNA in S phase subsequently converting to normal sized DNA.[156,157] This low molecular weight DNA corresponds to the distance between UV-induced pyrimidine dimers and reflects the inability of the cell's replicating polymerase to bypass these lesions.[158]

DNA replication studies in normal cell free extracts have shown that when a replication fork meets a lesion on the lagging strand template, DNA synthesis stops.[159,160] The replication fork continues to move along the template until it reaches the next Okazaki fragment and DNA synthesis reinitiates.[161] A small single-stranded gap remains that is resynthesized by pol δ or ε. When the lesion is in the leading strand, the replication fork is blocked temporarily. The lesion can be bypassed by template-strand switching. In this case, replication of the leading and lagging strands are uncoupled. The lesion blocks DNA synthesis of the leading strand, while the lagging strand continues to be replicated.[162,163] The resulting single-stranded DNA has the same polarity as the damaged parental strand and is a good substrate for transient template switching by DNA polymerase δ or ε. DNA polymerase δ bypasses the lesion by copying the daughter strand and then switching back to its original template.[165] Another mechanism for bypass of a lesion in the template strand is translesion synthesis (TLS).[166] This is a two-step process.[167] A specialized polymerase inserts nucleotides opposite the damaged bases in the template strand. This is followed by a short extension before resumption of replication fork progression.[168]

Evidence that the PRR deficiency in XPV is due to a deficiency in TLS came from replication studies in cell free extracts.[169] Unlike cell free extracts from HeLa cells, extracts from XPV cells were unable to sustain efficient progression of a replication fork past a *cis-syn* thymine-thymine CPD placed in the leading strand template. The addition of a purified protein from HeLa cells complemented replication activity past a CPD in XPV extracts.[170] The protein was sequenced, its cDNA counterpoint cloned and identified as a DNA polymerase homologous to the yeast DNA polymerase ε, a member of the Y-family of DNA polymerases[171] involved in translesion DNA synthesis of replication blocking lesions. Its coding gene, Pol E, has been shown to be mutated in all XPV affected patients examined.[172,173] Most mutations are found in the N-terminal region and abolish DNA polymerase and lesion bypass activities.[174]

Different Y-family members have different specificities for TLS bypass of different types of lesions.[175,176] The characteristics of the Y family of polymerases are large substrate binding sites and relaxed substrate specificity, resulting in low fidelity.[171,177] Since damaged bases lose their coding capacity, a nucleotide inserted across from a lesion is likely to result in a base substitution. If the nucleotide inserted were complementary to the next template base, a single base deletion would generate a frameshift mutation. In fact, their error rate for misinsertions and single base deletions on undamaged DNA is greater than 1%.[178-180] As a result, replication tends to be error prone but TLS past different types of DNA damage is possible. The Y family polymerases also show low processivity allowing them to quickly dissociate from their template after inserting only a few bases.[181,182] This allows the higher fidelity polymerase δ to reengage in replication immediately after TLS bypass.

Polymerase ε is uniquely proficient in accurately bypassing a *cis-syn* T-T dimer, but cannot fully bypass a 6-4 photoproduct.[183] Pol ε incorporates a G opposite the 3' T base of a T-T 6-4 PP and does not bypass the 5' base.[178,181] It has been hypothesized that a polymerase switch occurs and pol ζ a member of the classical B-family, completes TLS bypass of the 5' T and the short extension step.[184] This misincorporation of a G opposite a T would result in a T to C transition. However, 6-4 PP's are rapidly removed by NER within a few hours following UV irradiation.[185] In this case, TLS capability would be counterproductive. On the other hand, CPDs are removed slowly by NER and may remain in the genome through replication.[186] If so, they would act to arrest replication and result in either genomic instability or cell death. Polymerase η TLS would allow accurate DNA replication past these lesions.

As expected, fibroblasts from XPV patients are hypersensitive to UV-induced replication inhibition.[187] After UV irradiation, XPV cells display a high mutation rate consistent with the patients' high incidence of skin cancers.[188,189] Examination of p53 gene mutations in XPV patients' skin tumors has proven that the UV mutational spectrum differs from that of normal individuals and XPC patients.[190] In XPV skin tumors, 39% of the mutations were CC to TT transitions and 22% were C to T transitions at dipyrimidine sites.[67] The mutations found in normal individuals were mainly C to T transitions at dipyrimidine sites in contrast to the majority of mutations in XPC skin cancers being CC to TT transitions.[67,191]

Polymerase ε complementation studies proved that, in XPV cells, it is a deficiency of polymerase ε that is responsible for the UV hypermutability phenotype as well as the shift from a normal UV mutational spectrum.[192-193] In XPV, the highest levels of mutations were found at TC and CC sites as is expected for the UV mutational spectrum.[192] Additionally, there were high levels of errors at TT sites demonstrating that other DNA polymerases can inaccurately bypass TT lesions.[192] The functional complementation of XPV cells by pol ε lowered the number of errors during replication past UV lesions at TT and at a 3' C of a dipyrimidine site.[192] It is clear that the presence of pol ε offers some degree of protection against UV damage that escapes the NER pathway.

Multifunctional DNA Repair Proteins

Some of the repair proteins involved in NER have additional functions in other cellular processes. A link was established between basal transcription and NER when XPB was found to be the largest protein subunit of TFIIH.[194] Subsequently, XPD was determined to be the second largest protein subunit of TFIIH.[195] TFIIH is composed of nine subunits, which can be found in two sub complexes. The core TFIIH sub complex contains the XPB, p62, p52, p44, and p34 proteins.[196] The cdk-activating kinase (CAK) complex contains cdk7, cyclin H, and MAT1. The XPD protein is found associated with both of these complexes and brings them together.[197-198] TFIIH DNA unwinding activity is required for both transcriptional initiation and promoter escape as well as for DNA repair.[82] Both XPB and XPD are helicases. The XPB protein unwinds DNA in the 3' to 5' direction.[194] The XPD protein unwinds DNA in the 5'-3' direction.[199] Functional XPB helicase activity is required for both DNA repair and transcription.[200] On the other hand, XPD helicase activity is essential for DNA repair and not for transcription.[201] One would expect that mutations that eliminate the helicase activity of XPD would result in NER impairment but have little effect on transcription. However, XPD also acts to stabilize the TFIIH complex by binding to p44 in the core complex and by interacting with the CAK subcomplex.[201] Mutations in XPD that impair its binding to p44 result in reduced phosphorylation of several transcriptional activators by the cdk7 kinase subunit of TFIIH. These transactivators include the nuclear hormone receptors, RARα, RARγ, and Erα. Mutations in the XPD gene have been shown to result in decreased phosphorylation of nuclear receptors by TFIIH and a concomitant drop in expression of their target genes.[203] Such an impairment of hormonal responsive genes could explain the developmental abnormalities associated with the more severe cases of XPD such as XPD/CS.

Two other proteins, XPF and ERCC1, have important incision roles outside of NER. The XPF-ERCC1 complex is involved in cleavage of interstrand DNA cross links,[204] processing nonhomologous ends during recombination[205] and maintaining telomere integrity.[206] The complex associates with TRF2 and is present at telomere where it removes the G-strand overhangs from uncapped telomeres. Most importantly, it is believed to cleave recombination intermediates formed between telomeres and interstitial telomere-like sequences. Deficiencies in levels of XPF-ERCC1 result in the appearance of Telomeric DNA-containing Double Minute chromosomes (TDMs), which are circular extrachromosomal elements.[206]

In addition to its incision function in NER, the XPG protein has been implicated in the removal of oxidative DNA damage.[207] Only 13 XPG cases have been identified.[208,209] XPG patients combined with severe CS have both *XPG* alleles encoding mRNAs containing a PTC that would produce severely truncated XPG protein and act as substrates for NMD. In the two patients examined, XPG mRNA was barely detectable by Northern Blot analysis.[208] These patients would have no functional XPG protein. Other patients with mild XP symptoms were compound heterozygotes with one allele containing a PTC and the other containing a single point mutation that encodes a full length protein with greatly reduced or abolished XPG endonuclease activity.[209] Oxidative residues such as thymine glycol and 8-oxoguanine are still removed by TCR to the same extent in XPG cells as in normal cells but not in XPG/CS cells.[210] Additionally, both normal and mutated XPG proteins stimulate base excision repair activity of NTH1, a DNA glycosylase-apurinic lyase, which removes oxidized pyrimidines in the total genome.[207] Likewise, CS cells have been shown to be impaired in TCR of oxidative damage.[210] It has been suggested that failure of RNA polymerase to bypass oxidative damage contributes to the CS phenotype.

Conclusion

It is clear that UV sensitivity and high cancer susceptibility can be explained by defective DNA repair. However, many mutations in DNA repair genes result in multisystem disorders. Mutations in the same gene can result in different clinical outcomes and severity depending on the site of a mutation and the gene dosage. This is further complicated by the possibility of pleiotropic effects caused by disturbances in other cellular processes. While not discussed here, attempts to recapitulate the human genotype-phenotype relationships using transgenic mice are providing valuable insights into these disorders. It may soon be possible to determine a patient's clinical prognosis by analyzing the site of the mutation in the affected gene.

References

1. Kaposi M. Xeroderma pigmentosum. Ann Dermatol Venereol 1883; 4:29-38.
2. Cockayne EA. Dwarfism with retinal atrophy and deafness. Archives of Disease in Childhood 1946; 21:52-54.
3. Cockayne EA. Dwarfism with retinal atropy and deafness. Archives of Disease in Childhood 1936; 11:1-8.
4. Robbins JH et al. Xeroderma pigmentosum. An inherited diseases with sun sensitivity, multiple cutaneous neoplasms, and abnormal DNA repair. Ann Intern Med 1974; 80(2):221-48.
5. Robbins JH. Xeroderma pigmentosum. Defective DNA repair causes skin cancer and neurodegeneration. JAMA 1988; 260(3):384-8.
6. Cleaver JE. Xeroderma pigmentosum: A human disease in which an initial stage of DNA repair is defective. Proc Natl Acad Sci USA 1969; 63(2):428-35.
7. Bootsma D et al. Nucleotide excision repair syndromes: Xeroderma Pigmentosum, cockayne syndrome, and Trichothiodystropy. In: Vogelstein B, Kinzler KW, eds. The Genetic Basis of Human Cancer. New York: McGraw-Hill, Health Professions Division, 1998:245-274.
8. Rapin I et al. Cockayne syndrome and xeroderma pigmentosum. Neurology 2000; 55(10):1442-9.
9. Kraemer KH. Sunlight and skin cancer: Another link revealed. Proc Natl Acad Sci USA 1997; 94(1):11-4.
10. Kraemer KH et al. The role of sunlight and DNA repair in melanoma and nonmelanoma skin cancer. The xeroderma pigmentosum paradigm. Arch Dermatol 1994; 130(8):1018-21.
11. Kraemer KH, Lee MM, Scotto J. Xeroderma pigmentosum. Cutaneous, ocular, and neurologic abnormalities in 830 published cases. Arch Dermatol 1987; 123(2):241-50.
12. Robbins JH et al. Neurological disease in xeroderma pigmentosum. Documentation of a late onset type of the juvenile onset form. Brain 1991; 114(Pt 3):1335-61.
13. Price VH et al. Trichothiodystrophy: Sulfur-deficient brittle hair as a marker for a neuroectodermal symptom complex. Arch Dermatol 1980; 116(12):1375-84.
14. Neill C, Dingwall M. A syndrome resembling progeria: A review of two cases. Archives of Disease in Childhood 1950; 25:213-223.
15. Macdonald WB, Fitch KD, Lewis.IC. Cockayne's syndrome. An heredo-familial disorder of growth and development. Pediatrics 1960; 25:997-1007.
16. Goldsmith LA. Genetic skin diseases with altered aging. Arch Dermatol 1997; 133(10):1293-5.
17. Patton MA et al. Early onset Cockayne's syndrome: Case reports with neuropathological and fibroblast studies. J Med Genet 1989; 26(3):154-9.
18. Lehmann AR, Norris PG. DNA repair and cancer: Speculations based on studies with xeroderma pigmentosum, Cockayne's syndrome and trichothiodystrophy. Carcinogenesis 1989; 10(8):1353-6.
19. Lehmann AR. Three complementation groups in Cockayne syndrome. Mutat Res 1982; 106(2):347-56.

20. Tanaka K et al. Genetic complementation groups in cockayne syndrome. Somatic Cell Genet 1981; 7(4):445-55.

21. Nance MA, Berry SA. Cockayne syndrome: Review of 140 cases. Am J Med Genet 1992; 42(1):68-84.

22. Lowry RB. Early onset of Cockayne syndrome. Am J Med Genet 1982; 13(2):209-10.

23. Moyer DB et al. Cockayne syndrome with early onset of manifestations. Am J Med Genet 1982;13(2):225-30.

24. Itoh T, Ono T, Yamaizumi M. A new UV-sensitive syndrome not belonging to any complementation groups of xeroderma pigmentosum or Cockayne syndrome: Siblings showing biochemical characteristics of Cockayne syndrome without typical clinical manifestations. Mutat Res 1994; 314(3):233-48.

25. Greenhaw GA et al. Xeroderma pigmentosum and Cockayne syndrome: Overlapping clinical and biochemical phenotypes. Am J Hum Genet 1992; 50(4):677-89.

26. Lindenbaum Y et al. Xeroderma pigmentosum/Cockayne syndrome complex: First neuropathological study and review of eight other cases. Eur J Paediatr Neurol 2001; 5(6):225-42.

27. Rasmussen RE, Painter RB. Evidence for repair of ultra-violet damaged deoxyribonucleic acid in cultured mammalian cells. Nature 1964; 203:1360-2.

28. Cleaver JE. Defective repair replication of DNA in xeroderma pigmentosum. Nature 1968; 218(142):652-6.

29. Cleaver JE. Xeroderma pigmentosum: Variants with normal DNA repair and normal sensitivity to ultraviolet light. J Invest Dermatol 1972; 58(3):124-8.

30. Collins AR. Mutant rodent cell lines sensitive to ultraviolet light, ionizing radiation and cross-linking agents: A comprehensive survey of genetic and biochemical characteristics. Mutat Res 1993; 293(2):99-118.

31. Thompson LH. Chinese hamster cells meet DNA repair: An entirely acceptable affair. Bioessays 1998; 20(7):589-97.

32. Westerveld A et al. Molecular cloning of a human DNA repair gene. Nature 1984; 310(5976):425-9.

33. Weber CA et al. Molecular cloning and biological characterization of a human gene, ERCC2, that corrects the nucleotide excision repair defect in CHO UV5 cells. Mol Cell Biol 1988; 8(3):1137-46.

34. Weeda G et al. Molecular cloning and biological characterization of the human excision repair gene ERCC-3. Mol Cell Biol 1990; 10(6):2570-81.

35. Thompson LH et al. Molecular cloning of the human nucleotide-excision-repair gene ERCC4. Proc Natl Acad Sci USA 1994; 91(15):6855-9.

36. Mudgett JS, MacInnes MA. Isolation of the functional human excision repair gene ERCC5 by intercosmid recombination. Genomics 1990; 8(4):623-33.

37. Troelstra C et al. Molecular cloning of the human DNA excision repair gene ERCC-6. Mol Cell Biol 1990; 10(11):5806-13.

38. Kirchner JM et al. Cloning and molecular characterization of the Chinese hamster ERCC2 nucleotide excision repair gene. Genomics 1994; 23(3):592-9.

39. O'Donovan A, Wood RD. Identical defects in DNA repair in xeroderma pigmentosum group G and rodent ERCC group 5. Nature 1993; 363(6425):185-8.

40. Legerski R, Peterson C. Expression cloning of a human DNA repair gene involved in xeroderma pigmentosum group C. Nature 1992; 360(6404):610.

41. Henning KA et al. The Cockayne syndrome group A gene encodes a WD repeat protein that interacts with CSB protein and a subunit of RNA polymerase II TFIIH. Cell 1995; 82(4):555-64.

42. Tanaka K et al. Molecular cloning of a mouse DNA repair gene that complements the defect of group-A xeroderma pigmentosum. Proc Natl Acad Sci USA 1989; 86(14):5512-6.

43. Chu G, Chang E. Xeroderma pigmentosum group E cells lack a nuclear factor that binds to damaged DNA. Science 1988; 242(4878):564-7.

44. Dualan R et al. Chromosomal localization and cDNA cloning of the genes (DDB1 and DDB2) for the p127 and p48 subunits of a human damage-specific DNA binding protein. Genomics 1995; 29(1):62-9.

45. Nichols AF, Ong P, Linn S. Mutations specific to the xeroderma pigmentosum group E Ddb- phenotype. J Biol Chem 1996; 271(40):24317-20.

46. Tanaka K et al. Analysis of a human DNA excision repair gene involved in group A xeroderma pigmentosum and containing a zinc-finger domain. Nature 1990; 348(6296):73-6.

47. Weeda G et al. Localization of the xeroderma pigmentosum group B-correcting gene ERCC3 to human chromosome 2q21. Genomics 1991; 10(4):1035-40.

48. Legerski RJ et al. Assignment of xeroderma pigmentosum group C (XPC) gene to chromosome 3p25. Genomics 1994; 21(1):266-9.

49. Flejter WL et al. Correction of xeroderma pigmentosum complementation group D mutant cell phenotypes by chromosome and gene transfer: Involvement of the human ERCC2 DNA repair gene. Proc Natl Acad Sci USA 1992; 89(1):261-5.

50. Sijbers AM et al. Xeroderma pigmentosum group F caused by a defect in a structure-specific DNA repair endonuclease. Cell 1996; 86(5):811-22.

51. Takahashi E, Shiomi N, Shiomi T. Precise localization of the excision repair gene, ERCC5, to human chromosome 13q32.3-q33.1 by direct R-banding fluorescence in situ hybridization. Jpn J Cancer Res 1992; 83(11):1117-9.

52. Yuasa M et al. Genomic structure, chromosomal localization and identification of mutations in the xeroderma pigmentosum variant (XPV) gene. Oncogene 2000; 19(41):4721-8.

53. Broughton BC et al. Two individuals with features of both xeroderma pigmentosum and trichothiodystrophy highlight the complexity of the clinical outcomes of mutations in the XPD gene. Hum Mol Genet 2001; 10(22):2539-47.

54. Lehmann AR. The xeroderma pigmentosum group D (XPD) gene: One gene, two functions, three diseases. Genes Dev 2001; 15(1):15-23.

55. Broughton BC et al. Molecular and cellular analysis of the DNA repair defect in a patient in xeroderma pigmentosum complementation group D who has the clinical features of xeroderma pigmentosum and Cockayne syndrome. Am J Hum Genet 1995; 56(1):167-74.

56. Stefanini M et al. Genetic analysis of twenty-two patients with Cockayne syndrome. Hum Genet 1996; 97(4):418-23.

57. Moriwaki S et al. DNA repair and ultraviolet mutagenesis in cells from a new patient with xeroderma pigmentosum group G and cockayne syndrome resemble xeroderma pigmentosum cells. J Invest Dermatol 1996; 107(4):647-53.

58. Matsui MS, DeLeo VA. Photocarcinogenesis by Ultraviolet A and B. In: Mukhtar H, ed. Skin Cancer: Mechanisms and Human Relevance. Boca Raton: CRC Press, 1995:21-30.

59. Tornaletti S, Pfeifer GP. UV damage and repair mechanisms in mammalian cells. Bioessays 1996; 18(3):221-8.

60. Kwa RE, Campana K, Moy RL. Biology of cutaneous squamous cell carcinoma. J Am Acad Dermatol 1992; 26(1):1-26.

61. Nataraj AJ, Trent II JC, Ananthaswamy HN. p53 gene mutations and photocarcinogenesis. Photochem Photobiol 1995; 62(2):218-30.

62. Dumaz N et al. Specific UV-induced mutation spectrum in the p53 gene of skin tumors from DNA-repair-deficient xeroderma pigmentosum patients. Proc Natl Acad Sci USA 1993; 90(22):10529-33.

63. Ziegler A et al. Sunburn and p53 in the onset of skin cancer. Nature 1994; 372(6508):773-6.

64. Campbell C et al. p53 mutations are common and early events that precede tumor invasion in squamous cell neoplasia of the skin. J Invest Dermatol 1993; 100(6):746-8.

65. Setlow RB. Cyclobutane-type pyrimidine dimers in polynucleotides. Science 1966; 153(734):379-86.

66. Sato M et al. Ultraviolet-specific mutations in p53 gene in skin tumors in xeroderma pigmentosum patients. Cancer Res 1993; 53(13):2944-6.

67. Giglia-Mari G, Sarasin A. TP53 mutations in human skin cancers. Hum Mutat 2003; 21(3):217-28.

68. Mullenders LH et al. UV-induced photolesions, their repair and mutations. Mutat Res 1993; 299(3-4):271-6.

69. Park H et al. Crystal structure of a DNA decamer containing a cis-syn thymine dimer. Proc Natl Acad Sci USA 2002; 99(25):15965-70.

70. Wacker A et al. Organic photochemistry of nucleic acids. Photochemistry and Photobiology 1964; 3:369-394.

71. Rycyna RE, Alderfer JL. UV irradiation of nucleic acids: Formation, purification and solution conformational analysis of the '6-4 lesion' of dTpdT. Nucleic Acids Res 1985; 13(16):5949-63.

72. Franklin WA, Haseltine WA. The role of the (6-4) photoproduct in ultraviolet light-induced transition mutations in E. coli. Mutat Res 1986; 165(1):1-7.

73. Kim JK, Choi BS. The solution structure of DNA duplex-decamer containing the (6-4) photoproduct of thymidylyl(3'—>5')thymidine by NMR and relaxation matrix refinement. Eur J Biochem 1995; 228(3):849-54.

74. Tornaletti S, Hanawalt PC. Effect of DNA lesions on transcription elongation. Biochimie 1999; 81(1-2):139-46.

75. Herzinger T et al. Ultraviolet B irradiation-induced G2 cell cycle arrest in human keratinocytes by inhibitory phosphorylation of the cdc2 cell cycle kinase. Oncogene 1995; 11(10):2151-6.

76. Svejstrup JQ. Mechanisms of transcription-coupled DNA repair. Nat Rev Mol Cell Biol 2002; 3(1):21-9.

77. Cleaver JE. Common pathways for ultraviolet skin carcinogenesis in the repair and replication defective groups of xeroderma pigmentosum. J Dermatol Sci 2000;23(1):1-11.

78. Lehmann AR. Dual functions of DNA repair genes: Molecular, cellular, and clinical implications. Bioessays 1998; 20(2):146-55.

79. Bohr VA et al. DNA repair in an active gene: Removal of pyrimidine dimers from the DHFR gene of CHO cells is much more efficient than in the genome overall. Cell 1985; 40(2):359-69.

80. Mellon I et al. Preferential DNA repair of an active gene in human cells. Proc Natl Acad Sci USA 1986; 83(23):8878-82.

81. Mellon I, Spivak G, Hanawalt PC. Selective removal of transcription-blocking DNA damage from the transcribed strand of the mammalian DHFR gene. Cell 1987; 51(2):241-9.

82. de Laat WL, Jaspers NG, Hoeijmakers JH. Molecular mechanism of nucleotide excision repair. Genes Dev 1999; 13(7):768-85.

83. Volker M et al. Sequential assembly of the nucleotide excision repair factors in vivo. Mol Cell 2001; 8(1):213-24.

84. Riedl T, Hanaoka F, Egly JM. The comings and goings of nucleotide excision repair factors on damaged DNA. EMBO J 2003; 22(19):5293-303.

85. Sugasawa K et al. Xeroderma pigmentosum group C protein complex is the initiator of global genome nucleotide excision repair. Mol Cell 1998; 2(2):223-32.

86. Christians FC, Hanawalt PC. Inhibition of transcription and strand-specific DNA repair by alpha-amanitin in Chinese hamster ovary cells. Mutat Res 1992; 274(2):93-101.

87. Conaway JW et al. Control of elongation by RNA polymerase II. Trends Biochem Sci 2000; 25(8):375-80.

88. Holstege FC, van der Vliet PC, Timmers HT. Opening of an RNA polymerase II promoter occurs in two distinct steps and requires the basal transcription factors IIE and IIH. EMBO J 1996; 15(7):1666-77.

89. van Hoffen A et al. Deficient repair of the transcribed strand of active genes in Cockayne's syndrome cells. Nucleic Acids Res 1993; 21(25):5890-5.

90. Venema J et al. The genetic defect in Cockayne syndrome is associated with a defect in repair of UV-induced DNA damage in transcriptionally active DNA. Proc Natl Acad Sci USA 1990; 87(12):4707-11.

91. Citterio E et al. ATP-dependent chromatin remodeling by the Cockayne syndrome B DNA repair-transcription-coupling factor. Mol Cell Biol 2000; 20(20):7643-53.

92. Bregman DB et al. UV-induced ubiquitination of RNA polymerase II: A novel modification deficient in Cockayne syndrome cells. Proc Natl Acad Sci USA 1996; 93(21):11586-90.

93. Tantin D, Kansal A, Carey M. Recruitment of the putative transcription-repair coupling factor CSB/ERCC6 to RNA polymerase II elongation complexes. Mol Cell Biol 1997; 17(12):6803-14.

94. van Gool AJ et al. The Cockayne syndrome B protein, involved in transcription-coupled DNA repair, resides in an RNA polymerase II-containing complex. EMBO J 1997; 16(19):5955-65.

95. Kamiuchi S et al. Translocation of Cockayne syndrome group A protein to the nuclear matrix: Possible relevance to transcription-coupled DNA repair. Proc Natl Acad Sci USA 2002; 99(1):201-6.

96. Neer EJ et al. The ancient regulatory-protein family of WD-repeat proteins. Nature 1994; 371(6495):297-300.

97. Nakatsu Y et al. XAB2, a novel tetratricopeptide repeat protein involved in transcription-coupled DNA repair and transcription. J Biol Chem 2000; 275(45):34931-7.

98. Venema J et al. Xeroderma pigmentosum complementation group C cells remove pyrimidine dimers selectively from the transcribed strand of active genes. Mol Cell Biol 1991; 11(8):4128-34.

99. Venema J et al. The residual repair capacity of xeroderma pigmentosum complementation group C fibroblasts is highly specific for transcriptionally active DNA. Nucleic Acids Res 1990; 18(3):443-8.

100. Tang JY et al. Xeroderma pigmentosum p48 gene enhances global genomic repair and suppresses UV-induced mutagenesis. Mol Cell 2000; 5(4):737-44.

101. Hanawalt PC, Ford JM, Lloyd DR. Functional characterization of global genomic DNA repair and its implications for cancer. Mutat Res 2003; 544(2-3):107-14.

102. Keeney S et al. Correction of the DNA repair defect in xeroderma pigmentosum group E by injection of a DNA damage-binding protein. Proc Natl Acad Sci USA 1994; 91(9):4053-6.

103. Rapic Otrin V et al. Relationship of the xeroderma pigmentosum group E DNA repair defect to the chromatin and DNA binding proteins UV-DDB and replication protein A. Mol Cell Biol 1998; 18(6):3182-90.

104. Hwang BJ et al. Expression of the p48 xeroderma pigmentosum gene is p53-dependent and is involved in global genomic repair. Proc Natl Acad Sci USA 1999; 96(2):424-8.

105. Hwang BJ, Chu G. Purification and characterization of a human protein that binds to damaged DNA. Biochemistry 1993; 32(6):1657-66.

106. Keeney S, Chang.GJ, Linn S. Characterization of a human DNA damage binding protein implicated in xeroderma pigmentosum E. J Biol Chem 1993; 268(28):21293-300.

107. Hwang BJ, Liao JC, Chu G. Isolation of a cDNA encoding a UV-damaged DNA binding factor defective in xeroderma pigmentosum group E cells. Mutat Res 1996; 362(1):105-17.

108. Hwang BJ et al. p48 Activates a UV-damaged-DNA binding factor and is defective in xeroderma pigmentosum group E cells that lack binding activity. Mol Cell Biol 1998; 18(7):4391-9.

109. Groisman R et al. The ubiquitin ligase activity in the DDB2 and CSA complexes is differentially regulated by the COP9 signalosome in response to DNA damage. Cell 2003; 113(3):357-67.

110. Fujiwara Y et al. Characterization of DNA recognition by the human UV-damaged DNA-binding protein. J Biol Chem 1999; 274(28):20027-33.

111. Aboussekhra A et al. Mammalian DNA nucleotide excision repair reconstituted with purified protein components. Cell 1995; 80(6):859-68.

112. Kazantsev A et al. Functional complementation of xeroderma pigmentosum complementation group E by replication protein A in an in vitro system. Proc Natl Acad Sci USA 1996; 93(10):5014-8.

113. Wakasugi M et al. DDB accumulates at DNA damage sites immediately after UV irradiation and directly stimulates nucleotide excision repair. J Biol Chem 2002; 277(3):1637-40.

114. Fitch ME et al. In vivo recruitment of XPC to UV-induced cyclobutane pyrimidine dimers by the DDB2 gene product. J Biol Chem 2003; 278(47):46906-10.

115. Tang J, Chu G. Xeroderma pigmentosum complementation group E and UV-damaged DNA-binding protein. DNA Repair (Amst) 2002; 1(8):601-16.

116. van Hoffen A et al. Transcription-coupled repair removes both cyclobutane pyrimidine dimers and 6-4 photoproducts with equal efficiency and in a sequential way from transcribed DNA in xeroderma pigmentosum group C fibroblasts. EMBO J 1995; 14(2):360-7.

117. Masutani C et al. Purification and cloning of a nucleotide excision repair complex involving the xeroderma pigmentosum group C protein and a human homologue of yeast RAD23. EMBO J 1994; 13(8):1831-43.

118. Araki M et al. Centrosome protein centrin 2/caltractin 1 is part of the xeroderma pigmentosum group C complex that initiates global genome nucleotide excision repair. J Biol Chem 2001; 276(22):18665-72.

119. Sugasawa K et al. A multistep damage recognition mechanism for global genomic nucleotide excision repair. Genes Dev 2001; 15(5):507-21.

120. Kusumoto R et al. Diversity of the damage recognition step in the global genomic nucleotide excision repair in vitro. Mutat Res 2001; 485(3):219-27.

121. Hey T et al. The XPC-HR23B complex displays high affinity and specificity for damaged DNA in a true-equilibrium fluorescence assay. Biochemistry 2002; 41(21):6583-7.

122. Lee JH et al. NMR structure of the DNA decamer duplex containing double T*G mismatches of cis-syn cyclobutane pyrimidine dimer: Implications for DNA damage recognition by the XPC-hHR23B complex. Nucleic Acids Res 2004; 32(8):2474-81.

123. Sugasawa K et al. A molecular mechanism for DNA damage recognition by the xeroderma pigmentosum group C protein complex. DNA Repair (Amst) 2002; 1(1):95-107.

124. Batty D et al. Stable binding of human XPC complex to irradiated DNA confers strong discrimination for damaged sites. J Mol Biol 2000; 300(2):275-90.

125. Tapias A et al. Ordered conformational changes in damaged DNA induced by nucleotide excision repair factors. J Biol Chem 2004; 279(18):19074-83.

126. de Laat WL et al. DNA-binding polarity of human replication protein A positions nucleases in nucleotide excision repair. Genes Dev 1998; 12(16):2598-609.

127. Li L et al. An interaction between the DNA repair factor XPA and replication protein A appears essential for nucleotide excision repair. Mol Cell Biol 1995; 15(10):5396-402.

128. He Z et al. RPA involvement in the damage-recognition and incision steps of nucleotide excision repair. Nature 1995; 374(6522):566-9.

129. Brookman KW et al. ERCC4 (XPF) encodes a human nucleotide excision repair protein with eukaryotic recombination homologs. Mol Cell Biol 1996; 16(11):6553-62.

130. Gary R et al. The DNA repair endonuclease XPG binds to proliferating cell nuclear antigen (PCNA) and shares sequence elements with the PCNA-binding regions of FEN-1 and cyclin-dependent kinase inhibitor p21. J Biol Chem 1997; 272(39):24522-9.

131. Kaiser MW et al. A comparison of eubacterial and archaeal structure specific 5'-exonucleases. J Biol Chem 1999; 274(30):21387-94.

132. Kao HI et al. Cleavage specificity of Saccharomyces cerevisiae flap endonuclease 1 suggests a double-flap structure as the cellular substrate. J Biol Chem 2002; 277(17):14379-89.

133. Hohl M et al. Structural determinants for substrate binding and catalysis by the structurespecific endonuclease XPG. J Biol Chem 2003; 278(21):19500-8.

134. Evans E et al. Open complex formation around a lesion during nucleotide excision repair provides a structure for cleavage by human XPG protein. EMBO J 1997; 16(3):625-38.

135. Miura M et al. Roles of XPG and XPF/ERCC1 endonucleases in UV-induced immunostaining of PCNA in fibroblasts. Exp Cell Res 1996; 226(1):126-32.

136. Park CH et al. Purification and characterization of the XPF-ERCC1 complex of human DNA repair excision nuclease. J Biol Chem 1995; 270(39):22657-60.

137. de Laat WL et al. DNA structural elements required for ERCC1-XPF endonuclease activity. J Biol Chem 1998; 273(14):7835-42.

138. Kuraoka I et al. Repair of an interstrand DNA cross-link initiated by ERCC1-XPF repair/recombination nuclease. J Biol Chem 2000; 275(34):26632-6.

139. Mu, D et al. Characterization of reaction intermediates of human excision repair nuclease. J Biol Chem 1997; 272(46):28971-9.

140. Huang JC et al. Human nucleotide excision nuclease removes thymine dimers from DNA by incising the 22nd phosphodiester bond 5' and the 6th phosphodiester bond 3' to the photodimer. Proc Natl Acad Sci USA 1992; 89(8):3664-8.

141. Shivji KK, Kenny MK, Wood RD. Proliferating cell nuclear antigen is required for DNA excision repair. Cell 1992; 69(2):367-74.

142. Krishna TS et al. Crystal structure of the eukaryotic DNA polymerase processivity factor PCNA. Cell 1994; 79(7):1233-43.

143. Schurtenberger P et al. The solution structure of functionally active human proliferating cell nuclear antigen determined by small-angle neutron scattering. J Mol Biol 1998; 275(1):123-32.

144. Ducoux M et al. Mediation of proliferating cell nuclear antigen (PCNA)-dependent DNA replication through a conserved p21(Cip1)-like PCNA-binding motif present in the third subunit of human DNA polymerase delta. J Biol Chem 2001; 276(52):49258-66.

145. Eissenberg JC et al. Mutations in yeast proliferating cell nuclear antigen define distinct sites for interaction with DNA polymerase delta and DNA polymerase epsilon. Mol Cell Biol 1997; 17(11):6367-78.

146. Maga G, Hubscher U. Proliferating cell nuclear antigen (PCNA): A dancer with many partners. J Cell Sci 2003; 116(Pt 15):3051-60.

147. Shiomi Y et al. ATP-dependent structural change of the eukaryotic clamp-loader protein, replication factor C. Proc Natl Acad Sci USA 2000; 97(26):14127-32.

148. O'Donnell M, Jeruzalmi D, Kuriyan J. Clamp loader structure predicts the architecture of DNA polymerase III holoenzyme and RFC. Curr Biol 2001; 11(22):R935-46.
149. Wood RD, Shivji MK. Which DNA polymerases are used for DNA-repair in eukaryotes? Carcinogenesis 1997; 18(4):605-10.
150. Shivji MK et al. Nucleotide excision repair DNA synthesis by DNA polymerase epsilon in the presence of PCNA, RFC, and RPA. Biochemistry 1995; 34(15):5011-7.
151. Karimi-Busheri F et al. Repair of DNA strand gaps and nicks containing 3'-phosphate and 5'-hydroxyl termini by purified mammalian enzymes. Nucleic Acids Res 1998; 26(19):4395-400.
152. Friedberg EC, Walker GC, Siede W. Nucleotide excision repair: Mammalian genes and proteins. In DNA Repair and Mutagenesis. Washington, DC: ASM Press, 1995:352-353.
153. Hofmann H, Jung EG, Schnyder UW. Pigmented Xerodermoid: First report of a family. Bull Cancer 1978; 65(3):347-50.
154. Ensch-Simon I, Burgers PM, Taylor JS. Bypass of a site-specific cis-Syn thymine dimer in an SV40 vector during in vitro replication by HeLa and XPV cell-free extracts. Biochemistry 1998; 37(22):8218-26.
155. Svoboda DL, Briley LP, Vos JM. Defective bypass replication of a leading strand cyclobutane thymine dimer in xeroderma pigmentosum variant cell extracts. Cancer Res 1998; 58(11):2445-8.
156. di Caprio L, Cox BS. DNA synthesis in UV-irradiated yeast. Mutat Res 1981; 82(1):69-85.
157. Prakash L. Characterization of postreplication repair in Saccharomyces cerevisiae and effects of rad6, rad18, rev3 and rad52 mutations. Mol Gen Genet 1981; 184(3):471-8.
158. Lehmann AR. Postreplication repair of DNA in ultraviolet-irradiated mammalian cells. J Mol Biol 1972; 66(3):319-37.
159. Svoboda DL, Vos JM. Differential replication of a single, UV-induced lesion in the leading or lagging strand by a human cell extract: Fork uncoupling or gap formation. Proc Natl Acad Sci USA 1995; 92(26):11975-9.
160. Nikolaishvili-Feinberg N, Cordeiro-Stone M. Bypass replication in vitro of UV-induced photoproducts blocking leading or lagging strand synthesis. Biochemistry 2001; 40(50):15215-23.
161. Meneghini R, Hanawalt P. T4-endonuclease V-sensitive sites in DNA from ultraviolet-irradiated human cells. Biochim Biophys Acta 1976; 425(4):428-37.
162. Cordeiro-Stone M et al. Analysis of DNA replication forks encountering a pyrimidine dimer in the template to the leading strand. J Mol Biol 1999; 289(5):1207-18.
163. Cordeiro-Stone M et al. Replication fork bypass of a pyrimidine dimer blocking leading strand DNA synthesis. J Biol Chem 1997; 272(21):13945-54.
164. Higgins NP, Kato K, Strauss B. A model for replication repair in mammalian cells. J Mol Biol 1976; 101(3):417-25.
165. Cordonnier AM, Fuchs RP. Replication of damaged DNA: Molecular defect in xeroderma pigmentosum variant cells. Mutat Res 1999; 435(2):111-9.
166. Pages V, Fuchs RP. How DNA lesions are turned into mutations within cells? Oncogene 2002; 21(58):8957-66.
167. Baynton K, Fuchs RP. Lesions in DNA: Hurdles for polymerases. Trends Biochem Sci 2000; 25(2):74-9.
168. McCulloch SD, Kokoska RJ, Kunkel.TA. Efficiency, fidelity and enzymatic switching during translesion DNA synthesis. Cell Cycle 2004; 3(5):580-3.
169. Cordonnier AM, Lehmann AR, Fuchs RP. Impaired translesion synthesis in xeroderma pigmentosum variant extracts. Mol Cell Biol 1999; 19(3):2206-11.
170. Masutani C et al. Xeroderma pigmentosum variant (XP-V) correcting protein from HeLa cells has a thymine dimer bypass DNA polymerase activity. EMBO J 1999; 18(12):3491-501.
171. Ohmori H et al. The Y-family of DNA polymerases. Mol Cell 2001; 8(1):7-8.
172. Johnson RE et al. hRAD30 mutations in the variant form of xeroderma pigmentosum. Science 1999; 285(5425):263-5.
173. Masutani C et al. The XPV (xeroderma pigmentosum variant) gene encodes human DNA polymerase eta. Nature 1999; 399(6737):700-4.
174. Kannouche P et al. Domain structure, localization, and function of DNA polymerase eta, defective in xeroderma pigmentosum variant cells. Genes Dev 2001; 15(2):158-72.

175. Rattray AJ, Strathern JN. Error-prone DNA polymerases: When making a mistake is the only way to get ahead. Annu Rev Genet 2003; 37:31-66.
176. Yang W. Damage repair DNA polymerases Y. Curr Opin Struct Biol 2003; 13(1):23-30.
177. Trincao J et al. Structure of the catalytic core of S. cerevisiae DNA polymerase eta: Implications for translesion DNA synthesis. Mol Cell 2001; 8(2):417-26.
178. Johnson RE et al. Fidelity of human DNA polymerase eta. J Biol Chem 2000; 275(11):7447-50.
179. Matsuda T et al. Low fidelity DNA synthesis by human DNA polymerase-eta. Nature 2000; 404(6781):1011-3.
180. Matsuda T et al. Error rate and specificity of human and murine DNA polymerase eta. J Mol Biol 2001; 312(2):335-46.
181. Masutani C et al. Mechanisms of accurate translesion synthesis by human DNA polymerase eta. EMBO J 2000; 19(12):3100-9.
182. McCulloch SD et al. Preferential cis-syn thymine dimer bypass by DNA polymerase eta occurs with biased fidelity. Nature 2004; 428(6978):97-100.
183. Washington MT et al. Accuracy of lesion bypass by yeast and human DNA polymerase eta. Proc Natl Acad Sci USA 2001; 98(15):8355-60.
184. Johnson RE et al. Role of DNA polymerase zeta in the bypass of a (6-4) TT photoproduct. Mol Cell Biol 2001; 21(10):3558-63.
185. Carty MP, Lawrence CW, Dixon K. Complete replication of plasmid DNA containing a single UV-induced lesion in human cell extracts. J Biol Chem 1996; 271(16):9637-47.
186. Friedberg EC, Walker GC, Siede W. DNA Repair and Mutagenesis. Washington, DC: ASM Press, 1995.
187. Fischer E, Jung EG, Cleaver JE. Pigmented xerodermoid and XP-variants. Arch Dermatol Res 1980; 269(3):329-30.
188. Wang YC et al. Evidence from mutation spectra that the UV hypermutability of xeroderma pigmentosum variant cells reflects abnormal, error-prone replication on a template containing photoproducts. Mol Cell Biol 1993; 13(7):4276-83.
189. McGregor WG et al. Abnormal, error-prone bypass of photoproducts by xeroderma pigmentosum variant cell extracts results in extreme strand bias for the kinds of mutations induced by UV light. Mol Cell Biol 1999; 19(1):147-54.
190. Spatz A et al. Association between DNA repair-deficiency and high level of p53 mutations in melanoma of Xeroderma pigmentosum. Cancer Res 2001; 61(6): 2480-6.
191. Sarasin A. The molecular pathways of ultraviolet-induced carcinogenesis. Mutat Res 1999; 428(1-2):5-10.
192. Stary A et al. Role of DNA polymerase eta in the UV mutation spectrum in human cells. J Biol Chem 2003; 278(21):18767-75.
193. Yamada A et al. Complementation of defective translesion synthesis and UV light sensitivity in xeroderma pigmentosum variant cells by human and mouse DNA polymerase eta. Nucleic Acids Res 2000; 28(13):2473-80.
194. Schaeffer L et al. DNA repair helicase: A component of BTF2 (TFIIH) basic transcription factor. Science 1993; 260(5104):58-63.
195. Schaeffer L et al. The ERCC2/DNA repair protein is associated with the class II BTF2/TFIIH transcription factor. EMBO J 1994; 13(10):2388-92.
196. Winkler GS et al. Affinity purification of human DNA repair/transcription factor TFIIH using epitope-tagged xeroderma pigmentosum B protein. J Biol Chem 1998; 273(2):1092-8.
197. Reardon JT et al. Isolation and characterization of two human transcription factor IIH (TFIIH)-related complexes: ERCC2/CAK and TFIIH. Proc Natl Acad Sci USA 1996; 93(13):6482-7.
198. Drapkin R et al. Human cyclin-dependent kinase-activating kinase exists in three distinct complexes. Proc Natl Acad Sci USA 1996; 93(13):6488-93.
199. Roy R et al. The DNA-dependent ATPase activity associated with the class II basic transcription factor BTF2/TFIIH. J Biol Chem 1994; 269(13):9826-32.
200. Tirode F et al. Reconstitution of the transcription factor TFIIH: Assignment of functions for the three enzymatic subunits, XPB, XPD, and cdk7. Mol Cell 1999; 3(1):87-95.
201. Winkler GS et al. TFIIH with inactive XPD helicase functions in transcription initiation but is defective in DNA repair. J Biol Chem 2000; 275(6):4258-66.

202. Coin F et al. Mutations in the XPD helicase gene result in XP and TTD phenotypes, preventing interaction between XPD and the p44 subunit of TFIIH. Nat Genet 1998; 20(2):184-8.
203. Keriel A et al. XPD mutations prevent TFIIH-dependent transactivation by nuclear receptors and phosphorylation of RARalpha. Cell 2002; 109(1):125-35.
204. Sasaki MS et al. Recombination repair pathway in the maintenance of chromosomal integrity against DNA interstrand crosslinks. Cytogenet Genome Res 2004; 104(1-4):28-34.
205. Adair GM et al. Role of ERCC1 in removal of long nonhomologous tails during targeted homologous recombination. EMBO J 2000; 19(20):5552-61.
206. Zhu XD et al. ERCC1/XPF removes the 3' overhang from uncapped telomeres and represses formation of telomeric DNA-containing double minute chromosomes. Mol Cell 2003; 12(6):1489-98.
207. Klungland A et al. Base excision repair of oxidative DNA damage activated by XPG protein. Mol Cell 1999; 3(1):33-42.
208. Emmert S et al. Relationship of neurologic degeneration to genotype in three xeroderma pigmentosum group G patients. J Invest Dermatol 2002; 118(6):972-82.
209. Lalle P et al. The founding members of xeroderma pigmentosum group G produce XPG protein with severely impaired endonuclease activity. J Invest Dermatol 2002; 118(2):344-51.
210. Le Page F et al. Transcription-coupled repair of 8-oxoguanine: Requirement for XPG, TFIIH, and CSB and implications for Cockayne syndrome. Cell 2000; 101(2):159-71.

Index